INNOVATION, PATENTS AND TECHNOLOGICAL STRATEGIES

ORGA ... OPMENT

ORGANISATION FOR ECONOMIC CO-OPERATION AND DEVELOPMENT

Pursuant to Article 1 of the Convention signed in Paris on 14th December 1960, and which came into force on 30th September 1961, the Organisation for Economic Co-operation and Development (OECD) shall promote policies designed:

- to achieve the highest sustainable economic growth and employment and a rising standard of living in Member countries, while maintaining financial stability, and thus to contribute to the development of the world economy;
- to contribute to sound economic expansion in Member as well as non-member countries in the process of economic development; and
- to contribute to the expansion of world trade on a multilateral, non-discriminatory basis in accordance with international obligations.

The original Member countries of the OECD are Austria, Belgium, Canada, Denmark, France, Germany, Greece, Iceland, Ireland, Italy, Luxembourg, the Netherlands, Norway, Portugal, Spain, Sweden, Switzerland, Turkey, the United Kingdom and the United States. The following countries became Members subsequently through accession at the dates indicated hereafter: Japan (28th April 1964), Finland (28th January 1969), Australia (7th June 1971), New Zealand (29th May 1973), Mexico (18th May 1994) and the Czech Republic (21st December 1995). The Commission of the European Communities takes part in the work of the OECD (Article 13 of the OECD Convention).

Publié en français sous le titre :
INNOVATION, BREVETS ET STRATÉGIES TECHNOLOGIQUES

FOREWORD

The papers presented in this publication were among those contributed by participants from a wide variety of backgrounds at the "Innovation, Patents and Technological Strategies" workshop held at OECD headquarters in Paris on 8-9 December 1994. The workshop was arranged to bring together the preliminary results of new innovation surveys, as well as recent studies using patent data to reveal strategies in specific industrial or technological sectors.

Discussions focused on three main topics: industrial structures, innovation, and patents; company innovation and patenting strategies; globalisation of technology and patents. The papers selected reflect these three topics and show the present situation as it appeared at the workshop. At the same time, they discuss applications of the new methods of analysis, their complementarity and limitations, and suggest improvements or new areas for research.

The opinions expressed in this volume are not necessarily those of the Organisation or Member countries. It is published on the responsibility of the Secretary-General of the OECD.

TABLE OF CONTENTS

INTRODUCTION

BACKGROUND

It has long been known that R&D indicators give an incomplete picture of technological change in industry. Given the difficulties in measuring R&D output, patents are now being widely used as an indirect indicator of the effectiveness of the means devoted to innovation and of the technological performance of firms, industries and countries.

The patent systems of industrialised countries were originally intended for protecting the intellectual property rights of firms and individuals for the products or processes they invented; they can also act as an incentive for further innovation. Nevertheless, the use of patents as performance indicators for innovation has serious limitations, which have been pointed out in numerous studies on innovation and highlighted in the OECD Patent Manual.[1]

Because of the limitations of these indicators, users have been prompted to collect information on technological activity through innovation surveys. Recognition of the extent to which the results of innovation were "appropriated" resulted in the Yale[2] and PACE[3] surveys and the inclusion of specific questions on this subject in innovation surveys. At the same time, progress was made in analysing patent data used for technological indicators.

For most of these studies, the data collection phase is almost finished and the initial findings are being studied. The workshop on Innovation, Patents and Technological Strategies, organised by the OECD Directorate for Science, Technology and Industry on 8-9 December 1994 in Paris, the results of which are presented in this publication, aimed to review past studies and work in progress in order to learn more about the structure and characteristics of industrial activity with respect to patents and innovations, the links between this activity, and the main influences on the technological strategies of firms. The information gained, which may interest analysts and policy makers, should bring out not only the relative significance of the various categories of data used to assess the technological profiles of firms and industry but also the fields in which more extensive expert studies should be conducted. It should also help to focus future OECD work on innovation indicators in particular and firms' technological performance in general.

In science and technology, the OECD acts as a clearing-house and has run a number of multi-stage programmes on methodology, surveys and definition of internationally comparable S&T indicators, amongst other things. This work has paved the way for empirical studies that are both useful to policy makers and conducive to a better understanding of the innovation process.

Before describing the main issues addressed during the workshop and the contributions to it, it is useful to trace the progress of OECD work on industrial technology indicators over the couple of decades separating the two major international exercises known as "Gaps in Technology" (1967-68) and the "Technology/Economy Programme" (TEP, 1988-90).[4]

FROM THE "GAPS IN TECHNOLOGY" STUDY TO THE TEP

The starting point for OECD work on measuring and analysing national performances in the area of technological innovation dates back to the "Gaps in Technology" operation launched in 1968 at the third "gaps in technology" meeting of Ministers for Science from OECD Member countries.

From the early 1950s onwards, the enormous progress made by US enterprises in certain high-tech sectors such as electronics had considerable political implications. By the early 1960s the outlay on research and development in the United States – as a percentage of GDP – was over double that of the main European countries and Japan (3.4 per cent compared to 1.5 per cent and 1.4 per cent, respectively, in 1964). It was against this background that the OECD decided to start work on developing internationally comparable indicators to gauge the structure and intensity of scientific and technological activity in Member countries prior to indicating the nature and extent of technology gaps – especially at sectoral level – and analysing their impact on performance.

The field covered was extremely broad, since it included not only R&D but also the three areas of scientific and technical personnel, technological innovation and international trade. In terms of R&D, the "Gaps in Technology" study was able to draw on the findings of the first OECD international survey to use the common methodology described in the *Frascati Manual,*[5] first published in 1963 (and since revised four times, the fifth edition appearing in 1994). At the time, the findings of this R&D survey represented a breakthrough, prompting Professor C. Freeman to declare that, until the survey, policy makers had had more precise information on national egg production than on the size of their research force. The review of technological innovation, for its part, was based on patents, the technology balance of payments, and some information on "technical" innovation.

The comparison between the United States, Europe and Japan relied on analysis of nine industrial sectors. US enterprises had a clear lead in the field of electrical engineering and electronics (calculators, semiconductors and scientific instruments) while in other sectors, such as chemicals, pharmaceuticals, man-made fibres, machinery, iron and steel, and ferrous metals, their advantage was generally less significant.

The conclusions of the General Report of the 1968 meeting on gaps in technology seem to see the difference between the United States on one hand and Europe and Japan on the other as a difference in the use of scientific, technological and industrial capabilities rather than a difference in levels of S&T resources in the United States on one hand and Europe and Japan on the other. The concept of a "gap in technology" is linked to that of a "gap in management techniques" for the innovation process.

In other words, the need to define common principles of national science policy had been recognised in the early 1960s and was now being officially reaffirmed and extended to technology.

On the basis of the work on technology gaps, it was possible to test the findings of international surveys in OECD countries on resources earmarked for R&D. In doing so, their limitations for analysing the technological level of an industry or country were revealed, and this encouraged the production of other S&T data.

OECD work on improving the measurement of technological innovation and output indicators (in relation to the input indicators represented by the human and financial resources channelled into R&D) dates back to the late 1970s. At international level such work only really began following experts' recommendations in 1978 and 1979. These recommendations were confirmed at an OECD conference in 1980, the objective of which was to achieve a consensus between indicator users and producers on the output indicators considered most promising for international comparisons.

It was only in 1990 that the first methodological document of the *Frascati* family of manuals to measure R&D output rather than input was published. It dealt with the technology balance of payments (TBP), which is used in association with other indicators to evaluate the technological competitiveness of national economies. It was also the first practical guide written for producers and users of data on international trade in disembodied technology. The document in question is the "Proposed Standard Method of Compiling and Interpreting Technology Balance of Payments Data – TBP Manual".[6]

In the absence of directly measured data on technological innovation, patent statistics have been used as a source of information on invention activities and on the diffusion of technology and innovation. They were used intensively for these purposes in the 1980s owing to the expansion of the European Patent Office (EPO) and the boom in computerised databases. Since they combine three separate aspects (legal, technical and economic), they are very complex and must be interpreted with care. The "1994 Patent Manual" proposes ways of approaching the various problems, encourages the use of patent data, and advocates international harmonisation of patent statistics.

However, patents do not necessarily represent innovation, and there is an urgent need for direct measurements in order to achieve a better understanding of the innovation process and to obtain a range of more varied information.

"The OECD Proposed Guidelines for Collecting and Interpreting Technological Innovation Data – Oslo Manual",[7] based on the experience of national experts and a conceptual framework proposed by the Nordic countries, was adopted in the autumn of 1991. At the instigation of the European Commission, this collection of proposed guidelines served as a guide for the design of the first series of harmonised surveys (Community Innovation Survey – CIS) which were carried out in the European Union and a number of OECD Member countries.

In order to draw on experience from these initial surveys, especially with regard to analysis of conceptual and methodological problems that may give rise to difficulties in interpretation or international comparison, a joint OECD/European Commission review of the Oslo Manual has recently started.

Lastly, the Technology/Economy Programme, launched by the OECD Council in 1988 with the main aim of working towards an integrated and global approach to technological, economic and social issues, has provided an opportunity to assess the requirements for quantitative data in order to improve the latter as well as to take advantage of new analytical tools and conceptual frameworks. Consequently it has made the S&T indicators of the DSTI move forward in three respects:

- It changed the conceptual framework on which all OECD statistical work on S&T indicators was based, since it helped to challenge the traditional idea of R&D as the hinge linking science and innovation in a linear innovation process and replaced it with a more systemic concept which stresses the interactive processes at work in developing, disseminating and improving knowledge.
- By urging the need for a range of indicators, the TEP prompted the drafting and distribution of the above-mentioned Frascati family of manuals which deal with measurement of S&T output and impact. The latest addition is the manual on the measurement of human resources devoted to S&T ("Canberra Manual"),[8] produced jointly by the OECD and Eurostat.
- The TEP gave a decisive boost to work on using S&T and industrial-performance indicators for structural analysis in fields such as technology dissemination and trends in international competitiveness. This work has resulted in important new databases.[9]

In the wake of the "Gaps in Technology" exercise, the Technology/Economy Programme thus demonstrated that our understanding of the links between science and technology on the one hand, and economic performance and well-being on the other – a precondition of policy making – is heavily dependent on the availability of appropriate internationally comparable statistics and indicators.

CONTRIBUTIONS TO THE "INNOVATION, PATENTS AND TECHNOLOGICAL STRATEGIES" WORKSHOP

This publication is a collection of several articles selected from those contributed by national experts who attended the workshop on "Innovation, Patents and Technological Strategies" held at OECD headquarters on 8-9 December 1994. As indicated by its title, the workshop had three objectives, which it pursued by referring to recent studies based on patent statistics as well as the preliminary results of new technological innovation surveys.

- To achieve a better understanding of the impact of industrial structures and firm size on the level of innovation and patent activity in a country and to check the consistency of results obtained from different sources.
- To identify the main determinants of firms' strategies with regard to innovation and patents and to indicate the relative importance of the various mechanisms of innovation appropriation and different protection mechanisms used.
- To endeavour to trace the international forms taken by technological development within major firms (co-operation agreements, alliances, take-overs, etc.) as well as the division of labour between the parent company and its subsidiaries.

D. Archibugi and M. Pianta begin with a presentation of the principal conceptual and methodological aspects encountered by those investigating innovative activities in depth. They then review firms' innovation and patenting strategies, the importance of industrial structure, and innovative activity at national level and the process of globalisation. Different points of view are presented on the new intellectual property rules and on how the benefits of innovation are appropriated by firms and countries. By way of conclusion, the authors summarise the results of the workshop on patent-based and technological innovation indicators, the main methodological problems encountered, the research themes discussed; finally, they propose new avenues for future research.

M. Åkerblom, M. Virtaharju and A. Leppälahti examine, for Finland, the results of innovation surveys, R&D surveys and studies based on patent statistics using a statistical key in order to compare the Finnish industrial classification and the International Patent Classification (IPC).

C. Grenzmann and S. Greif jointly carried out a detailed technical study to gain insights into the behaviour of German firms regarding R&D investment and patenting activities. The analysis was made possible by a full-scale census identifying, at this initial stage, 5 100 firms. Their paper presents the results of this work and outlines the projected follow-up.

S. Lhuillery's paper is a review of the preliminary results of the French survey carried out in 1993 which was partially based on the Community Innovation Survey questionnaire. Two issues were addressed: a new performance variance representing the capacity to remain in the innovation process, and the means used by enterprises to acquire technology, especially the R&D links it forges with different partners.

P. Kaminski looks into micro-enterprises, which play an essential role in the process of creating and diffusing innovations between major firms and SMIs. The results of the article stem from the TIME survey carried out in autumn 1992 and spring 1993 in partnership with INSEE, the *Institut Supérieur des Métiers*, and Eurostat. The author highlights the features of very small firms, which differed from SMIs mainly in their undifferentiated operational structure (production, sales, market approach), which in fact largely depend on the personality of the head of the business. Similarly, the R&D activities of micro-enterprises do not lend themselves to the agreed definitions of the *Frascati Manual*. Micro-enterprises feel that any exchange of technical knowledge and know-how or any very close collaboration would imply expansion, which they see as a form of diminution or a loss of independence. Hence the need for government assistance or incentives specifically for small-scale industries.

Using data from the Italian innovation survey of 8 220 manufacturing firms, R. Evangelista studies the links, at enterprise and industry level, between disembodied innovative activities (measured in terms of R&D and design engineering expenditure per employee) and tangible investments in innovation (measured in terms of innovatory tools and equipment). The author notes that these investments are the principal component of innovative activities in the manufacturing industry as a whole and that, at the level of industry in general, disembodied innovative activities predominate in the general engineering, electrical and electronic industries; in the chemical, food, rubber and metal industries, instead, tangible investments in innovation prevailed. In aerospace, office equipment, telecommunications and radio and television sets, fairly high levels of R&D and investment in advanced technology go hand in hand.

S. Arvanitis and H. Hollenstein apply an empirical model in order to analyse the significance of determinants of innovative activities using data from an innovation survey of Swiss manufacturing industries carried out in 1993 by the *École Polytechnique Fédérale of Zurich*. In order to assess the validity of hypotheses on the propensity of enterprises to innovate, four groups of explanatory variables were used (outlook for demand, market conditions, appropriability and technological possibilities); however, it is evident that the dominant variables are those linked to supply (appropriability, technological possibilities).

C. DeBresson and M. Xiaoping take the results of recent surveys on innovative activities in Italy, France and China to try to establish and evaluate "innovative matrices" along the lines of Leontief's model of input-output tables in order to reveal each country's specialisations. In all three countries, they discovered that innovative activities form clusters in the sectors of economic activity having the greatest number of links upstream and downstream. The clusters form in different sections of the economy, depending on the country concerned, which would seem to indicate that there are differences at the inter-industrial systems level.

R. Barré attempts to assess the impact on national innovation systems (NIS) in 11 Member countries of the technology strategy of multinational enterprises, both in the country of origin and in the host country. Four methods of incorporating NIS into the international innovation networks of multinationals are identified; these methods vary greatly, reflecting the nature of the different NIS.

U. Schmoch first looks at patent flows between North America, Western Europe and Japan by contrasting the different strategies adopted by these three partners. This is followed by a study of how some 20 firms in the telecommunications sector made initial patent filings in various foreign countries from 1987 to 1989. In the case of telecommunications, the author notes a clear difference between international patent filing strategies and essentially national R&D policies, though the R&D activities of a number of firms are now much more international.

W.H. Brown and M.J. Hirabayashi, working from the TAF and USPTO patent databases of the United States, observe an increasing trend for the United States and other countries to co-operate in developing patentable technologies. Several features are studied, such as the inventor's country of residence, the patent-holder (enterprise, government, individual), and the technical domain. They demonstrate the significant role of subsidiaries in the make-up of international teams of inventors. They also identify the methodological problems outstanding and point out data discrepancies which hinder the satisfactory resolution of patent ownership issues.

H. Grupp, G. Münt and U. Schmoch focus on ways of determining which database is the most appropriate for analysing internationally relevant technology likely to have a great impact on international trade. This study, carried out by the Fraunhofer Institute for Systems and Innovation Research in Germany, culminated in a new international database called the "triad patent model". The authors claim that satisfactory results were achieved when this database was used to test the links between international technological activities and trade flows between the three economic blocs (North America, Western Europe and Japan).

CONCLUSION

The DSTI's work on science, technology and industry indicators and the conferences it has organised have thrown new light on the nature of technological change and its effects on productivity, competitiveness and employment. From these studies and discussions, it is apparent that the actions taken to intensify the role of technical change are global in nature, encompassing a wide range of institutional, economic, social and cultural components, and that the most effective analytical approach is to devise a national system of technology development.

It is therefore essential to diversify further the scope of information if we are to clarify the impact of S&T resources and of the innovation process in particular. This has become increasingly urgent over the past ten years, as innovation policy decisions have been made on very limited R&D- or patent-based statistics as the only regular sources of information.

With regard to the direct measurement of technological innovation – the theme of most of the articles collected here – it is clear that the available information on the innovation process is very varied and detailed. These items of information can usefully be compared by national policy makers with other sources of information they may have at hand, such as R&D and patent data. At present, the information applies to technological innovations only and is often limited to the country concerned. Major problems subsist regarding comparisons among countries, and it will be up to the national experts to decide which aspects it would be desirable and feasible to develop for future surveys and national studies on technological innovation.

NOTES AND REFERENCES

1. OECD (1994), "Using Patent Data as Science and Technology Indicators. Patent Manual", OCDE/GD(94)114, Paris.
2. LEVIN, R.C., A.K. KLEVORICK, R.R. NELSON and S.G. WINTER (1987), *Appropriating the Returns from Industrial Research and Development*, Brookings Papers on Economic Activity, pp. 783-820.
3. ARUNDEL, A.G. VAN DE PAAL and L. SOETE (1995), *Innovation Strategies of Europe's Largest Industrial Firms: Results of the PACE Survey on Information Sources, Public Research, Protection of Innovation and Government Programmes*, MERIT, Maastricht.
4. OECD (1992), *Technology and the Economy: The Key Relationships*, Paris.
5. OECD (1993), *Proposed Standard Practice for Surveys of Research and Experimental Development – Frascati Manual*, Paris.
6. OECD (1990), "Proposed Standard Method of Compiling and Interpreting Technology Balance of Payments Data – TBP Manual", Paris.
7. OECD (1992), "OECD Guidelines for Collecting and Interpreting Technological Innovation Data – Oslo Manual", OCDE/GD(92)26, Paris.
8. OECD (1995), "Manual on the Measurement of Human Resources Devoted to S&T – Canberra Manual", OCDE/GD(95)77, Paris.
9. OECD (1995), *The OECD STAN Database for Industrial Analysis, 1974-93*, Paris and OECD (1995), *Research and Development Expenditure in Industry, 1973-92*, Paris

INNOVATION SURVEYS AND PATENTS AS TECHNOLOGY INDICATORS: THE STATE OF THE ART

by

Daniele Archibugi and Mario Pianta
ISRDS-CNR, Italy

1. INTRODUCTION

Innovation is a complex and heterogeneous process that is increasingly important in economic and social life. This overview focuses on innovative activities carried out by firms in order to survey how the innovation process and technology strategies can be described and investigated by using patent data and indicators derived from innovation surveys.

In order to understand the innovation process, detailed sources of information are needed; they make it possible to develop appropriate indicators and strategies at both firm and government level. In industry, innovation depends on a variety of activities ranging from formalised R&D to production engineering. It has been stressed that innovation is not a linear process going from R&D activities to the eventual commercialisation of products. On the contrary, the elements of innovation interact throughout the various stages to weave a complex web of relationships (see OECD, 1992*b*).

Three main aspects of industrial innovation deserve to be mentioned. First, technological change impinges on codified and tacit knowledge. Second, the sources of innovation may be either internal or external to the firm. Third, innovations can either be embodied in capital goods and products or disembodied, *i.e.* the know-how included in patents, licences, design, R&D activities, or embodied in skilled personnel.

These aspects already indicate the complex and heterogeneous nature of technological change. They show why it is very difficult to find measures that provide a satisfactory account of its dimensions, intensity, rate and direction. However, such measures are much needed for analytical and policy purposes.

Two main issues need to be addressed: first, the extent to which available indicators overlap or provide information on different aspects of science and technology systems; second, the extent to which indicators on the same activities provide similar answers. These issues can be summarised in two questions: Which indicator answers which question? Do different indicators provide the same results?

National and international governmental agencies, as well as private companies, have devoted a substantial amount of resources to producing information and statistical sources on innovation. The "family" of OECD manuals provides a wide array of conceptual and operational tools for developing and using the existing technological indicators and statistical sources. In particular, the Patent Manual (OECD, 1994) and the Oslo Manual on innovation surveys (OECD, 1992*a*) cover the methodological aspects and the state of the art in these fields (see also the Introduction to this volume). A final question emerging from this overview concerns the further development of methodological tools: What are the new efforts to be agreed upon in order to provide a common framework for expanding the information drawn from technology indicators?

The papers presented at the OECD workshop on "Innovation, Patents and Technological Strategies", held in Paris on 8-9 December 1994 – of which a number are included in this volume – offer some answers to these questions. They deal with a wide range of topics and provide up-to-date information on current research in OECD countries. Among the issues addressed are: the methodological problems of comparing different indicators of innovative activities; specific aspects of innovation in firms and industries; the international dimension of technology and its link to economic performance.

This overview first sets out the key concepts and the methodological problems of investigating innovative activities (Section 2). Next, it discusses innovation and patenting strategies of firms (Section 3) and the analysis of industrial structure (Section 4). The national level and the process of globalisation are then discussed (Section 5). Following that, it addresses the more specific issue of intellectual property systems and how the benefits of innovation are appropriated by firms and countries (Section 6). Finally, the conclusions bring together the evidence gathered for patent and innovation survey indicators; the key methodological problems and themes of research are reviewed in the light of the OECD workshop, and new direction for research are suggested (Section 7).

2. CONCEPTS AND METHODOLOGY

2.1. Two methods of gathering information on industrial innovation

Patents and innovation surveys are two ways to acquire information on the innovative activities of firms. Figure 1 presents a conceptual framework. A wide variety of innovative activities are carried out by firms, among which the introduction of product and process innovations and patenting are two specific activities that can be documented by innovation surveys and patent data. Using such

◆ Figure 1. ***A framework for the analysis of innovation and patenting activities***

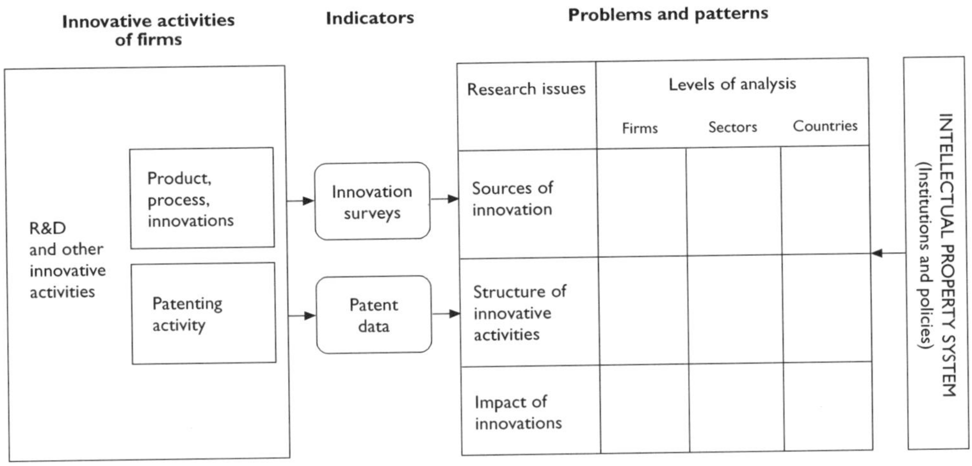

Source: Authors.

empirical evidence, it is possible to address several issues at the levels of firms, industries and countries.

Some innovation inputs have been monitored for a long time, notably the resources devoted to R&D, which have been systematically measured in most advanced countries for over 30 years. However, despite its importance, R&D is only one source of innovation. Other innovation inputs are not yet measured, and some cannot be (on indicators, see van Raan *et al.*, 1989; Barré *et al.*, 1994).

Innovative activities have a variety of visible outcomes, some of which are reflected in the performance of firms. Firms innovate in order to introduce new products or new processes into the market. Thus, one outcome of innovative activity is represented by the patents taken out by firms in order to protect their innovations against prospective competitors. The key aspects of patents and innovation surveys as technology indicators are shown in Table 1.

Innovation can be analysed, classified and measured from several perspectives. There are, at least, four different criteria for classifying innovative activities, which

Table 1. **The nature of patents and innovation surveys**

	Patents	Innovation surveys	
		"Object" approach	"Subject" approach
Unit of analysis	Patented inventions.	Sample of innovations.	Firms.
Origin of the information	Collected for legal and administrative purposes.	Collected for analytical and/or policy purposes.	Collected for analytical and/or policy purposes.
Method of collecting information	Patent office data on applications filed by inventors or grants. Original sources are often further reclassified and elaborated for analytical purposes.	Collected from different sources such as new product announcements, experts surveys, innovation inventories, bibliometric directories.	Collected at the firm level either by mail questionnaires or direct interviews.
Periodicity	Regular data collection. Very up-to-date information.	Occasional surveys.	Until now, occasional surveys. Data collection is becoming periodical.
Coverage	Inventions for which legal protection is sought. The majority of patents are granted to the business sector.	Samples of successful innovations. Informs on innovations introduced by both the business and the non-profit sectors.	Successful and unsuccessful innovative activities. Innovating and non-innovating firms. Includes manufacturing and service industries.
Main criteria of classification	Technological. Firm's principal economic activity.	Product. Firm's principal economic activity.	Firm's principal economic activity. Main user sector. Firm size.

Source: Author.

can be used in both patenting and innovation surveys (for a formal analysis, see Archibugi, 1988):

- technology, *i.e.* according to the technical characteristics of the innovation;
- product, *i.e.* according to the nature of the product in which the innovation is likely to be embodied;
- sector of production, *i.e.* the main economic activity of the firm that has generated the innovation;
- sector of use, *i.e.* the main economic activity of the users of the innovation.

2.2. The nature of patents

The patent system is one method firms use to protect their inventions. For legal reasons, patents are systematically registered by government bodies. If they are duly processed, classified and organised, patents provide a unique source of information on industrial innovation. The OECD Patent Manual provides guidelines for the use of patents as well as a guide to the patent-based literature.[1]

The nature of patent data and innovation surveys is summarised in Table 1; Table 2 provides an overview of their strengths and weaknesses. Like any other technological indicator, patents have advantages and disadvantages, which it is useful to summarise. Their advantages are:

- They are a direct outcome of the inventive process, and more specifically of those inventions which are expected to have a commercial impact. They are a particularly appropriate indicator for capturing the proprietary and competitive dimension of technological change.
- Because obtaining patent protection is time-consuming and costly, it is likely that applications are filed for those inventions which, on average, are expected to provide benefits that outweigh these costs.
- Patents are broken down by technical fields and thus provide information not only on the rate of inventive activity, but also on its direction.
- Patent statistics are available in large numbers and for a very long time series.

Their disadvantages are:

- Not all inventions are technically patentable. This is the case of software, which is generally legally protected by copyright.
- Not all inventions are patented. Firms sometimes protect their innovations with alternative methods, notably industrial secrecy.
- Firms have a different propensity to patent in their domestic market and in foreign countries, which largely depends on their expectations for exploiting their inventions commercially. In each national patent office, there are many more applications from domestic inventors than from foreigners.
- Although there are international patent agreements among most industrial countries, each national patent office has its own institutional characteristics, which affect the costs, length and effectiveness of the protection accorded. In turn, this affects the interest of inventors in applying for patent protection.

Empirical surveys suggest that a large share of firms' inventions are patented. Research carried out by Mansfield (1986) on a sample of US firms showed that firms apply for a patent for about 66 per cent to 87 per cent of their patentable inventions.

Table 2. **Comparability, strengths and weaknesses of patents and innovation surveys**

	Patents	Innovation surveys	
		"Object" approach	"Subject" approach
Time series comparability	Very high. Patent data have been collected for more than a century.	Generally high within a given survey.	Low, unless information is collected periodically and is standardised.
International comparability	High, although it is limited by the national nature of patent institutions and the large number of domestic applications.	Low. All surveys are national in scope. Difficult to compare them because of different sample method and design.	Potentially high for quantitative data if identical questionnaires and methods are used.
Comparability with R&D	Low at firm and industry levels. High at country level.	Low, since R&D surveys are at firm level and not at innovation level.	High, since surveys allow make it possible also to collect information on input data. Both innovation and R&D surveys collect information at firm level.
Comparability with industrial statistics and national accounts	Difficult at firm and industry levels due to different sectoral classifications. High at country level.	Low, because it is difficult or even impossible to relate the sampled innovation to the whole universe.	High on quantitative data if innovation surveys can be related to the economic universe.
Other advantages	• Since patent protection is costly, it is likely that firms apply for those inventions which are likely to provide economic returns. • Very detailed sectoral disaggregation.	• Direct measure of innovation. • Provides information on technological evolution.	• Provides information on all innovative activities. • Wide coverage of issues. • Informs on both producers and users of innovation.
Other disadvantages	• Not all patents become innovations. • Not all inventions are patented. • Not all inventions are patentable. • Different propensity to patent across sectors.	• Heterogeneous value of individual innovations. • Data biased by subjective judgement. • Difficult to assess the significance and representativeness of the sample.	• Does not inform on the technological nature of innovations. • Significance and representativeness of results are tied to response rate achieved.

Source: Author.

This does not mean that patents account for the same share of all inventions, since an unknown number of inventions are not technically patentable. Still, this suggests that firms have recourse to patenting for the majority of their patentable inventions.

A survey carried out by the European Patent Office (1994, p. 108) on the patenting activity of European firms with up to 1 000 employees found that 25 per cent of the sampled firms filed patent applications for more than 90 per cent of their patentable inventions and that another 25 per cent applied for a patent for 50 to 90 per cent of their patentable inventions. These shares tend to increase with the size of firms. Cost and lack of advantages are given as reasons for not patenting an invention.

A large-scale survey of firms' behaviour with respect to innovation and appropriability (the PACE survey) has recently been completed in Europe (Arundel *et al.*, 1995) and is under way in the United States and Japan. The survey was designed as a follow-up to the Yale survey of the 1980s (Levin *et al.*, 1987), and the first results broadly confirm previous evidence on firms' innovative strategies and on the relevance of patents.

The PACE survey of European firms found that 15 per cent of firms apply for patents for 80 to 100 per cent of their product innovations, while 37 per cent patent less than 19 per cent. For process innovations, only 7 per cent of firms made extensive use of patents and 57 per cent apply very rarely. The firms covered by the PACE survey, however, made extensive use of patents; only 14 per cent did not apply for a patent in the previous three years; 79 per cent applied at the EPO and 78 per cent at the national patent office; 66 per cent applied in the United States and 53 per cent in Japan (Arundel *et al.*, 1995, pp. 60-61).

A related question is whether patented inventions actually become innovations. Old (Scherer *et al.*, 1959) and new (Sirilli, 1987; Napolitano and Sirilli, 1990) empirical evidence shows very similar results: the share of patents actually used by firms ranges from 40 per cent to 60 per cent of total applications. Similarly, the EPO survey found that 47 per cent of European firms used commercially or licensed more than 90 per cent of their patented inventions, and another 16 per cent used between 50 and 90 per cent of their patents. These shares are higher for smaller firms and show substantial variations across countries and industries (EPO, 1994, pp. 117-119).

In the EPO survey, when firms were asked about their usual means of protecting new products and processes, 84 per cent of patenting firms cited patents in the case of products and 71 per cent in the case of processes. Secrecy was mentioned by about 50 per cent of patenting firms and by about 63 per cent of non-patenting firms. Getting to market ahead of competitors was reported by about 41 per cent of both patenting and non-patenting firms (EPO, 1994, pp. 88-91).

It has often been argued that the value of individual patents is highly skewed (see Swann, 1993). In fact, several methods have been used to assess the individual "quality" and "impact" of patents. Four different measures have been used to increase the accuracy of patent counts:

- *Patent citations*: the count of citations of a patent in subsequent patent literature. This is an indicator of the technological impact of the patented invention (see Narin and Olivastro, 1988; Trajtenberg, 1990).
- *Renewal fees*: the total cost and the number of years for which the patentee pays renewal fees to maintain the legal value of the patent. This informs on the economic value attributed to the invention (see Pakes and Simpson, 1989).
- *Patent families*: mapping the number of countries to which a single patent application has been extended makes it possible to identify the subset of patents applied for in all major markets. This shows the areas of exploitation of an invention and offers a more accurate database for international comparisons (for an application to the US, Japanese and the European Patent Offices, see Grupp, 1993; Schmoch and Kirsch, 1993).
- *Patent claims*: the number of claims made in each patent application, which informs on the range of novelties in the patent document.[2]

2.3. The nature of innovation surveys

Contrary to patent data, innovation surveys have been developed from the outset with the specific aim of acquiring information on innovative activities carried out in firms. Until very recently, innovation surveys were organised by government agencies, statistical offices or academic institutions for their own specific needs. In consequence, the results achieved differ quite significantly and are not easy to compare (for an attempt to do so, see Acs and Audretsch, 1991; Hansen, 1992; Smith, 1992*a*; Smith, 1992*b*; Kaminski, 1993). Tables 1 and 2 summarise the nature of the information drawn from innovation surveys and its strengths and weaknesses.

Innovation surveys have also to face up to the very heterogeneous nature of innovations. A jet engine, a microprocessor, but also a corkscrew or a hairpin, might all be classified as innovations. Several attempts have been made to classify innovations according to their economic and/or technological significance. Innovations have been divided into "improvement" versus "basic", "incremental" versus "discrete", "minor" versus "major", and Freeman (1992) has proposed a detailed taxonomy of innovations. Innovation surveys have also distinguished between innovations that are new at world level and those that are new for an individual country, industry or firm.

Recently, the OECD and other national and international organisations have made an effort to standardise the methodology and the information collected by innovation surveys. The Oslo Manual (OECD, 1992*a*) provides guidelines on the method and issues to be covered by innovation surveys. In collaboration with the OECD, Eurostat has prepared a harmonised questionnaire, and in 1993-94 the EU-sponsored Community Innovation Survey (CIS) was carried out. For the first time innovation surveys which are, to some extent, internationally comparable have been completed in most OECD Member countries (see Eurostat, 1994; Archibugi *et al.*, 1995). The first preliminary national findings to be published were those of France (see Chapter 4 in this volume), Italy (ISTAT, 1995), and Germany (Licht, 1994).

As Tables 1 and 2 indicate, there are two different approaches to innovation surveys. The first collects information at the level of individual innovations; this is the "object" approach. The information collected in this way is classified according to the "technological" or "product" criteria listed above (Section 2.1). The second collects information at the level of the firm producing and/or adopting the innovation; this is the "subject" approach. The information collected in this way is classified according to the "sector of production" of innovations as defined above (Section 2.1).

The "object" approach: innovation counts

In this approach, the individual innovation is the analytical unit of the survey. Additional information can also be collected and recorded, such as the size and main product line of the firm that introduced it. The object approach was adopted in order to acquire information on the dynamics of technological change in the context of the business cycle debate.

The SPRU innovation database, which contains information on 4 800 radical innovations in the United Kingdom since the Second World War (Townsend *et al.*, 1981) constituted a major development of this approach. The US Small Business Administration has collected information on 8 000 innovations commercialised in the United States in 1982 from technical and scientific journals and magazines (see Acs and Audretsch, 1990). A similar approach has also been used in some European countries (see Kleinknecht and Bain, 1993).[3]

The object approach has much in common with patent analysis since both represent innovation counts.[4] They aim to acquire information on the same target population, *i.e.* the innovations introduced into the economic system, although the methodology used to select the sampled observations is very different. While patents constitute a well-defined population – the inventions registered to secure legal protection – the same cannot be said for the population of innovation counts. In fact, none of the databases available claims to collect information either on all innovations introduced or on a statistically significant sample. In general, counts of innovations monitor fewer observations than patents, but record a larger amount of information for each of them.

The advantages of innovation surveys based on the object approach are:

- They represent a direct measure of innovation, and they only include innovations considered to be technologically and/or economically significant.
- They provide significant information on the evolution of technology, since they make it possible to record precisely when and how a certain innovation was introduced.

Their disadvantages are:

- The definition of the sample is arbitrary. Experts may have different perceptions of the relevance of individual innovations.
- It is very difficult to develop internationally comparable databases. Each of the surveys has used its own design, sample definition and implementation.

The "subject" approach: surveys of firms

This is an alternative method of acquiring direct information on innovation in industry, in which firms are surveyed to learn the inputs, outputs and characteristics of their innovative activities. While both patenting and innovation counts collect information on innovations *per se*, the subject approach also allows one to gather information on various topics related to innovative activities, on non-innovating firms, and on the factors that hamper innovation. It also allows one to collect information on innovative activities that do not lead to the introduction of actual innovations, because, for example, they resulted in failures.

Depending on the design of individual surveys, the nature and amount of information collected may vary considerably; only recently has an attempt been made to collect standardised information. The subject approach is part of the standard industrial economics statistics, since it collects, at firm level, information on both the inputs and outputs of innovative activities.

The Oslo Manual (Chapter 5, §137-140) has stressed the limits of the object approach, which are due to the heterogeneous nature of individual innovations. Following the Oslo Manual and the Community Innovation Survey launched in Europe, the subject approach is becoming the standard method for collecting direct information on innovation in industry.

Its main advantages are:

- The information collected can be related to the industrial structure. Data on innovation can be matched to economic data on production, value added, employment, etc., at the firm and industry level.

- It provides coverage of both innovating and non-innovating firms. This allows the factors preventing innovation to be explored.
- It gives information both on the firms generating and on those using innovations. This allows one to treat not only manufacturing but also the service industries.

Its main disadvantages are:

- In spite of recent efforts, it is not yet easy to gather internationally comparable data.
- Since the monitoring of innovation according to this method is still in its infancy, time-series comparisons are not yet possible.
- This method does not collect information on the technological nature of the innovations introduced in firms.

Considering the different advantages and disadvantages of technology indicators, such as patent data and variables derived from innovation surveys, it is important to compare the findings for a particular indicator to the picture provided by other data, including the more widely used R&D expenditure. Examples of such comparisons are offered in this volume. Thus, Chapter 2 of this volume compares R&D, total innovation expenditure, and sales according to innovative products and patents for Finland, and Chapter 3 compares R&D inputs and patent outputs for Germany. For France, a study of patents and innovation survey results for France can be found in Kabla (1994).

2.4. The industrial classifications of technology data

When comparing different technology indicators among themselves and/or to economic variables such as production and trade, the comparability of the classifications employed raises a serious problem. Innovation counts can provide different aggregations of the data collected, but a standardised classification has not been used so far. Innovation surveys of firms provide data at firm level which are consistent with the sources of industrial statistics. Provided that the survey is representative of the universe of firms, these data can be easily arranged so as to be comparable with information on production, value added, employment, etc., at all levels of aggregation, from individual firms to sectors of industrial activity.

Patent data are aggregated according to the International Patent Classification (IPC) and, in the United States, according to the US Patent Office classification (USPOC) for which a cross-classification concordance to the Standard Industrial Classification (SIC) has been established.

Economic indicators on production and trade are generally available from the International Standard Industrial Classification (ISIC), the Statistical Classification of Economic Activities in the European Community (NACE) and the Standard International Trade Classification (SITC). Depending on the type of analysis planned, different classifications have to be matched, and several efforts have been made in this direction.

Full information on the concordances with patent data that are available is provided in the Patent Manual (OECD, 1994). Patent data by SIC classes have been related to SITC data on trade (Grupp, 1991, 1992; Amendola *et al.*, 1992).

A concordance between IPC and ISIC classes has been developed at MERIT (Verspagen *et al.*, 1994) and a concordance between IPC and SITC has been developed for 46 high-technology product groups by ENEA, Cespri and Politecnico di Milano and is reported in Annex IV of the OECD Patent Manual. Other work has been carried out at FhG-ISI in Germany and at the Finnish Central Statistical Office.

3. COMPANY INNOVATION AND PATENTING STRATEGIES

After spelling out the key concepts and the methodological problems for the use of technology indicators, three major dimensions of analysis have to be addressed: the company level, the industry level, and the country level. These will be discussed in sequence, in terms of three major research topics: the sources, the structure, and the impact of innovations.

3.1. The sources of innovations for firms

Evidence from innovation surveys

The innovation surveys carried out so far (see, in particular, ISTAT, 1989; ministère de l'Industrie, 1994; and, for a review, Kaminski, 1993) have already highlighted several aspects of the sources, inputs and outputs of industrial innovation. One key piece of information provided by innovation surveys is the number of firms actually involved in innovation. Large firms or firms active in technologically dynamic sectors tend to innovate on a regular basis, while most companies, especially smaller ones, introduce innovations irregularly.

On such a critical question, however, the different design of surveys has so far prevented meaningful comparisons across countries. The framework provided by the OECD Oslo Manual and the Community Innovation Survey now makes possible the development of comparable evidence.

The CIS-based survey for Italy (ISTAT, 1995) covered close to 23 000 firms, one-third of which (7 500) have declared that they introduced innovations in the 1990-92 period. The share of innovating firms ranges from 84 per cent for firms with more than 1 000 employees to 26 per cent for firms with from 20 to 49 employees.

The CIS-based survey for France shows that in the same period close to 40 per cent of firms (using a sample of 4 500 firms) have declared that they introduced innovations, with similar differences across size groups (see Chapter 4 in this volume).

The CIS-based survey for Germany covers a sample of 2 900 firms; 65 per cent reported that they have introduced new or improved products or processes in the last three years or intend to do so in the coming years (Licht, 1994).

Innovation surveys have also produced new evidence on factors hampering innovation. Lack of funds, the excessive cost of innovation, and high risk are generally indicated as major obstacles (ISTAT, 1995). Such information can be useful in order to design specific policy measures destined either to increase the rate of innovation in firms that already introduce innovations or to overcome the barriers to innovation in those that have failed to do so.

Innovation surveys have also identified the sources used by firms to sustain innovation. According to the CIS-based surveys for Italy and France (ISTAT, 1995; Chapter 4 in this volume), sources internal to the firm (R&D and other departments), suppliers of equipment and materials, customers, exhibitions and trade fairs are (in that order) the main sources of innovations. Customers, technical journals, trade fairs and suppliers (in that order) emerge from the EPO survey as the main sources of information on technical developments for both patenting and non-patenting firms in Europe.

Another important finding on the sources of innovation is the fact of remarkable inter-industry differences. While traditional industries tend to use sources external to the firm and often acquire innovations embodied in capital goods, science-based industries use internal sources such as R&D and design (see Pavitt, 1984; von Hippel, 1988; Archibugi *et al.*, 1991).

Surveys have also investigated the organisation of R&D and innovative activities of firms. The results show a wide variety of patterns within firms, of links to other firms and other institutions, and of network relationships both within and across sectors.

A major contribution of innovation surveys is the information they provide on the various expenditures related to innovation. The CIS-based survey for Italy shows that R&D accounted for only 36 per cent of total innovation costs, with 14 per cent due to design and trial production, and 47 per cent devoted to innovation-related investment. The share of R&D grows with the size of firms, while the importance of investment falls (for firms with more than 1 000 employees, it is about 40 per cent) (ISTAT, 1995; the question was not included in the French survey).

These findings confirm previous evidence for the importance of non-R&D costs of innovation (see also Chapter 6 in this volume). However, each industry is characterised by a specific combination of embodied and disembodied, tangible and intangible sources of innovation.

Chapter 7 in this volume presents an attempt to explain firms' innovative activities. Using data from the Swiss innovation survey, the authors find that the degree of appropriability and technological opportunities affect firms' innovations more than demand or market conditions. Crepon and Duguet (1994) examine how market effects and externalities affect innovation in French firms and find a close relationship between R&D and patenting.

The multi-technology firm

Industrial economists and management analysts often need to identify the technological nature of the innovations developed within a firm. The majority of firms, and above all the large ones, are highly diversified, and this is reflected in their innovations. Their technological activities often cover a larger number of fields than their product lines. However, innovation surveys are not able to provide a satisfactory description of the technological nature of the innovations introduced.

Patents, instead, provide much more detailed information. The high level of disaggregation offered by patent data and the number of patents registered by large firms make it possible to investigate the distribution of a firm's innovative projects. Several studies have considered firms' patent portfolios either to study their technological diversification (Kodama, 1986; Niwa, 1992; Patel and Pavitt, 1994) or to identify to what extent firms benefit from innovations carried out by firms engaged in similar technological areas (Jaffe, 1986). These studies have shown that:

- the majority of companies have a wider distribution of technological activities than of product lines and that they often produce their own equipment and machinery, or the intermediate components of their products;
- patents can help to identify company strategies, often before they are implemented in the market;
- patents are also a valuable tool for identifying the combination of different branches of knowledge into a new technological advance (technology fusion).

3.2. The structure of innovation: firm size and concentration

A current important topic in industrial economics concerns how industrial concentration and market structure are affected by innovation.[5] What is the relative innovation performance of small and large firms? Is there a specific market structure (such as monopoly, oligopoly or perfect competition) that can maximise the rate of innovation?

Several indicators, and particular R&D and patenting, have been used to test these hypotheses. These indicators are particularly valuable for accounting for the innovations introduced in large firms. In fact, both R&D and patenting are concentrated in large firms: less than 700 companies are responsible for nearly 60 per cent of patenting in the United States, a share slightly higher than their share of business-funded R&D (see Patel and Pavitt, 1991; Casson, 1991). Similar levels of concentration also emerge at national level.

Neither R&D nor patents provide an accurate account of the distribution of innovative activities across firms of different sizes (see Ministère de l'Industrie, 1994). It has been shown that standard R&D surveys underestimate the amount of R&D carried out in small firms (Kleinknecht, 1987; Kleinknecht and Reijnen, 1991; Archibugi *et al.*, 1995).

However, beyond the question of the accuracy of the measurement of R&D activities, it appears that significant non-R&D innovative activities are carried out to a greater extent in smaller firms and in traditional industrial sectors. Innovation surveys have been used to assess the relative contribution of large and small firms to innovation (Pavitt *et al.*, 1987; Acs and Audretsch, 1990, 1992; Archibugi, Evangelista and Simonetti, 1995; Malerba and Orsenigo, 1995). The results of these studies have confirmed the existence of significant inter-industry differences in concentration ratios. They have also shown that small firms in selected fields are not disadvantaged in terms of innovation when compared to their larger competitors.

Other attempts have recently been made to identify innovation flows in specific industrial districts. It has been suggested that small firms tend to create their own networks for acquiring and transferring technical information. The specific role of very small firms in the innovation process and as part of supplier networks of large firm is discussed in Chapter 5 of this volume.

3.3. The impact on firms' performance

Patents and technological indicators also highlight the dynamics of the innovation process and its impact on the performance of firms. Technology is expected to influence firms' economic performance in a variety of ways, including productivity, growth and competitiveness. A major research project on these issues has been developed by the US National Bureau of Economic Research; it focuses on the relationship between technological factors (measured by R&D and patents) and economic indicators such as productivity and stock market value (see Griliches, 1984, 1990). It has shown that the technological performance of the firm is positively associated with its market value.

Using innovation counts, Acs and Audretsch (1990) have also studied the role of technology in the growth of firms. Simonetti (1994) has used patent indicators combined with other variables for US firms; Schwitalla and Grupp (1994) have undertaken a similar study for German firms. While these studies have confirmed that performance and technology are associated, they have also suggested that it is not easy to identify a general pattern of causality going from the latter variable to the former.

4. INDUSTRIAL STRUCTURE, INNOVATION, PATENTS

The characteristics of a firm's industrial structure play a key role in shaping its technological activities and performance. The intensity and scope of a firm's innovative efforts are strongly constrained by industry-specific aspects, including technological opportunities and market structure. Detailed information on the similarities and differences among industrial sectors and on flows of know-how is needed in order to understand the boundaries of firms' innovative activities.

As already pointed out above (Section 2.1), sectoral groupings of technological and industrial data can be developed on the basis of several criteria; attention should always be paid to the classifications used. As for firms, the role of industrial structure is here considered in terms of the sources, structure and impact of innovative activities.

4.1. Technological interdependence

Technology systems are characterised by strong interdependence. Some innovations might be produced and used within the same firm, but most significant innovations move among firms and sectors. Some innovations have a single user industry; others have a more pervasive impact across industries. Much progress has been made in understanding and measuring economic interdependence, using techniques such as input-output tables (for a review, see Archibugi, 1988; DeBresson, 1990). Acquiring information on technological interdependence requires the availability of data on innovations classified according to both sector of production and sector of use.

The study of technological interdependence has first a descriptive value. It makes possible, among other things, the identification of industries with strong interactions. It may also lead to important policy implications: identifying upstream suppliers of innovations for a specific industry may help design appropriate innovation policies.

Innovation flows across sectors

A variety of innovation surveys have also recorded and classified the industries that use innovations (see Robson *et al.*, 1988; DeBresson *et al.*, 1994). Most studies have been carried out at the country level and identify the specific characteristics of user-producer interactions. Some key intersectoral links have emerged as persistent patterns. Chapter 8 in this volume finds important differences in the industry clusters according to intensity of innovation flows in Italy, France and China, thereby confirming the different innovative and industrial structures of the three countries.

Patent flows across sectors

Information from user sectors can also be obtained from patent data. Each patent document is assigned to a sector according to its technical nature, but since the invention involved should also be "useful", a patent may also include a reference to its prospective use. The latter information, unfortunately, is not systematically collected in patent documents and databases.

Schmookler (1966) and Scherer (1982*b*) have identified the sector of use of individual patents, thereby making it possible to produce matrixes of technological interdependence, with each cell containing the number of patents that share the same industry of production and industry of use. From 1972, the Canadian patent office has also provided information on the industry of likely use of patents (see Englander *et al.*, 1988). Using Canadian data for the 1986-89 period, Hanel (1993) has produced a matrix for the production and use of inventions by 29 industries (the matrix is reported in the OECD Patent Manual, Annex IV-B).

4.2. The industrial structure of innovation

At the industrial level, it is possible to identify several common aspects of the structure of the innovative process. These have been summarised in various taxonomies based on different R&D intensities, on sources of innovations, on mechanisms of technical change, and on firms' strategies. Pavitt (1984) proposed a most successful taxonomy identifying four broad groups: industries where the introduction of innovations is generally based on R&D (science-based); industries where innovations largely come from suppliers of inputs and machinery (supplier-dominated); industries which emerge as specialised suppliers of innovative goods; industries where innovation and process technology are closely linked to scale factors (scale-intensive). This taxonomy has been widely applied to technological data, such as innovation counts and patents, either at industry level (Archibugi *et al.*, 1991; Kristensen, 1993) or at firm level (Cesaratto and Mangano, 1992), as well as to economic data on production and trade.

Chapter 6 in this volume develops further qualifications of industrial patterns using data from the first Italian innovation survey and considering firm size and the nature of innovative activities (disembodied know-how such as R&D and design and engineering on the one hand, and technology embodied in tangible innovative investment on the other). Four groups are identified: technology users that introduce innovations developed by other industries; small-scale innovators that emphasize innovative design and engineering; industries dominated by large-scale firms, which rely on investment in innovative machinery; sectors combining both high R&D and high tangible investment efforts.

4.3. The impact on the performance of industries

The distribution of a firm's sales according to the type of product innovation introduced is an important indicator of the economic impact of innovations. Such data are now included in the CIS-based innovation surveys; data for Italy, based on an extended questionnaire, show that 27 per cent of firms' sales were due to products with (radical or incremental) innovations; 28 per cent were affected by process innovations; and 45 per cent had no innovative content. While little change is found in groups of firms of different sizes, large differences emerge among sectors; 75 per cent of sales in the office machinery industry include product innovations, and shares of around 50 per cent are found for most machinery and electrical sectors, while values below 20 per cent are found in some traditional industries (ISTAT, 1995).

This evidence can be related to EPO survey findings that, among European firms applying for patents, 47 per cent of sales were unaffected by innovation, while 26 per cent were due to new or substantially changed products and processes and 27 per cent to improved products and processes. Among non-patenting firms, the share of sales with no innovation stood at 52 per cent. No differences by firm size have emerged in these patterns (EPO, 1994, p.74).

In the French CIS-based survey, 41 per cent of firms reported a share of innovative sales between 0 and 10 per cent of total sales and less than 20 per cent of firms reported that more than 30 per cent of their turnover was related to product innovations (see Chapter 4 in this volume).

This indicator of the market impact of innovations, as measured by the shares of sales of innovative products, offers a rather different picture from that of other technology indicators. In a preliminary analysis of the findings of the German CIS-based survey, the shares of innovative sales appeared largely unrelated to R&D and patent indicators (Licht, 1994). This confirms the results of the first Italian innovation survey (Cesaratto and Mangano, 1992).

In assessing the impact of innovation on industry performance, it should be borne in mind that the most innovative sectors are also those whose weight in the economy is increasing. However, as pointed out above (Section 4.1), a major result of innovations is a reduction in the (relative) costs of new products and activities and, therefore, generalised productivity gains in several user sectors. To account for such increases in productivity, innovative activities, measured either by patent or innovation surveys, should be classified for both sector of production and sector of use (see Scherer, 1982*a*; Englander *et al.*, 1988; Sterlacchini, 1989).

Electronics is a major example of an industry that is an important producer of innovations. The dramatic reduction in the cost of many electronic products makes a comparison between high technological activity (as measured by patents) and the value of production in the industry (with growing volumes, but falling prices) potentially misleading, as the benefits of the innovations are actually distributed to user industries.

An industry's ability to exploit the benefits of the innovations it has produced is strictly related to appropriability. If appropriability and market power are strong, all benefits can remain with the innovating firm, and the performance of the innovating firm shows productivity gains. Conversely, when appropriability is low and diffusion mechanisms and competition are strong, the benefits can quickly spread to users and consumers in the form of lower prices and/or better products or processes. In this case, productivity gains are expected to be found economy-wide; this is the assumption underlying research on innovation and "total factor productivity" in neo-classical models (see Griliches, 1984; Englander *et al.*, 1988).

Innovation surveys provide additional information on the impact on performance expected or experienced by the innovating firms. While most of this evidence is likely to be qualitative and difficult to relate in a systematic way to quantitative variables, new light can be shed, especially at the sectoral level, on the link between innovation and performance.

The impact innovations introduced by firms have had on employment and capital intensities offers a relevant example. In the first Italian innovation survey, carried out on 8 220 firms (see ISTAT 1989), it was asked whether the innovations introduced had led to an increase, decrease or no change in the use of labour and capital. A study at industry level by Vivarelli *et al.* (1995) shows the labour-substituting effects of innovations (in most sectors and sizes of firms) and their capital-deepening nature (in practically all cases). The overall negative impact of innovation on employment in Italian industry is found to be caused by the dominant role of process innovations and embodied technical change. A labour-increasing impact is only found in few sectors characterised by higher expenditures for design and engineering and higher shares of product innovations.

5. NATIONAL PATTERNS AND THE GLOBALISATION OF TECHNOLOGY

Recent research has emphasized the importance of national systems of innovation and their differences across OECD countries in terms of institutions, relationships among actors, size of resources, sectors of specialisation, and type of performance. In parallel, a growing literature has addressed the globalisation of technology, which mainly results from the activity of multinational firms. This section discusses the national patterns emerging from innovation and patent data and the impact of cross-border flows of technology on countries' position.

5.1. The global sources of innovation

The innovation process is increasingly taking place on a global scale. Archibugi and Michie (1995) have identified three major forms of globalisation of technology (Table 3).

First, the *global exploitation of technology* includes the use by firms of patents and other intellectual property rights to protect their inventions and block competitors as they prepare their entrance into foreign markets or licence their technology to local producers. This large and growing phenomenon (6 per cent growth rate a year) is due both to the practice of extending protection to more countries (three to four countries on average) and to the actual growth in the number of patented inventions. Evidence on this trend is also provided by the joint report of the patent offices of Europe, the United States and Japan (EPO, USPTO and JPO, 1993) and by a study by Schmoch and Kirsch (1993).

Table 3. **Three meanings of "techno-globalism"**

	Economic equivalent	Measure(s)	Results	
			Stock	Flow
***a)* Global exploitation of technology**	International trade flows (as opposed to foreign direct investment).	Patents extended in foreign markets. Technological balance of payments.	Patents were on average extended in 3-4 foreign markets in 1990.	6 per cent average annual growth rate for the OECD countries during the 1980s.
***b)* Global technological collaboration**	International joint-ventures.	Inter-firm technical agreements. Patent licencing.	Not available.	6 per cent annual growth comparing the period 1985-89 to the 1980-84 periods.
***c)* Global generation of technology**	Foreign direct investment (as opposed to trade flows).	Patents and R&D of firms from outside their home country.	3.8 per cent of US patents in 1981-86.	1 per cent growth between 1981-85 and 1986-90.

Source: Archibugi and Michie, 1995.

Second, the amount of *international collaboration* in firms' innovative efforts points to the pooling of resources in different countries, due to the search for complementarities in firms' technological and marketing strategies. It may involve not only to the generation of innovations but also their application, diffusion and adaptation to local markets. According to the database on inter-firm co-operation agreements developed at MERIT by Hagedoorn and Schankeraad (1993), there has been an average annual increase of 6 per cent in the number of international technology agreements between the first and the second half of the 1980s. International collaborations are revealed also in the rapid growth of patents with inventors from more than one country. Chapter 11 in this volume examines this aspect and shows that the number of US patents with at least one American and one foreign inventor grew from 90 in 1983 to 1 500 in 1993. An analysis of EPO patents with inventors from France and other countries is carried out in Duguet (1994).

Third, *global generation of technology* is found within single multinational firms when innovation is the result of efforts undertaken in laboratories and plants situated abroad (Pearce and Singh, 1992). An indicator of this tendency is provided by the share of patents granted to the foreign subsidiaries of multinational firms. Recent studies (Patel and Pavitt, 1991; Patel, 1995) have shown that these accounted for less than 4 per cent of total patents in the period 1981-86, with 1 per cent annual growth between the early and late 1980s. Chapter 10 in this volume presents similar results for multinational firms in the telecommunications sector.

These results suggest that the degree of globalisation in the generation of technology by large firms is lower than it is for production and investment, and that is confined to specific industries, countries and firms. The production of technology remains largely in (or in the vicinity of) the home country of the innovating firm.

While the globalisation process is important for selected firms and industries, it appears to have little influence on the overall picture of innovative activities that emerges from country data. Indeed, the characteristics of countries and of their national systems of innovation, namely their industrial strengths and fields of excellence, remain important for shaping the direction taken by the international flows of innovative activities and the strategies of multinational firms. Chapter 9 in this volume argues, however, that international innovative networks organised by multinational firms and national systems of innovation affect each other in a broader way, with a variety of country and sectoral specificities.

These results, however, hold mainly for the group of more advanced OECD countries that are active participants in the globalisation of technology. The perspective is different for the less advanced countries, which are mainly recipients of technology flows from abroad and which are more dependent on the activity of multinational firms. In terms of international flows of technology, the strategies of multinational firms and the policies of individual countries often diverge. This has become an important policy issue and is the object of economic negotiations among nations, for example in the recently completed GATT treaty (see Section 6).

5.2. The structure of innovative activities across countries

When the technological efforts of firms are investigated at country level, it is possible to identify the structure of innovative activities and, in particular, the breakdown by industry. Many studies have compared industrial patterns of innovation across countries in order to describe a country's relative technological position and its relative specialisations.

While the lack of internationally comparable innovation data has prevented cross-country analyses, patent data have been extensively used to investigate national strengths and weaknesses in technological fields, at different levels of aggregation. This is a rather straightforward application of patent data; however, attention should be paid to the nature of the databases and indicators used when making comparisons across countries, industries and over time (these issues are reviewed in detail in the OECD Patent Manual).

A large number of studies have provided a detailed description of countries' innovative activities and comparative performances and their changes over time. The use of different databases (patents in individual countries, in the United States, at the European Patent Office, in all triad countries, etc.) has made it possible to compare different pictures and to assess the value and limitations of patents as a technology indicator.

This evidence has shown that countries differ in their sectors of strengths and weaknesses in technology. Moreover, some countries concentrate their activity in selected fields, while others distribute their efforts more uniformly acoss technological areas. As expected, small countries show higher levels of specialisation in technological activities as well as in industrial productin and trade. Inspite of the relative convergence of aggregated technological efforts in advanced countries, the differences in their sectoral specialisation have generally increased (see, among others, Patel and Pavitt, 1987; Archibugi and Pianta, 1992).

5.3. Innovation and national performance

In order to investigate the impact technology has on the performance of countries and industries, several studies have related patent data to economic indicators, either at national level or when investigating industrial patterns.

Aggregate studies of the role of technology as a source of countries' competitiveness have shown that a higher intensity of technological activities has a generally positive impact on national growth. A large body of literature on technology and growth has investigated this relationship, using all the available indicators.

Patent data, disaggregated by product/industry, have been used to explore the relationship between technology and trade. Several studies have shown that sectoral specialisation resulting from patents is generally associated with the industrial pattern of countries' exports (Soete, 1987; Fagerberg, 1987, 1988; Cantwell, 1989; Dosi *et al.*, 1990; Amendola *et al.*, 1993; Verspagen, 1993). Other studies have explicitly addressed the problem of the concordance between different classifications discussed above (Section 2.4), developing a more accurate correspondence between patent classes and international trade categories (Soete, 1981: Amendola *et al.*, 1992). Chapter 12 in this volume links patent data and export performance of most OECD countries using a product-based classification of high technology.

Research on the patterns defined by the relationship between industries' innovative activity and national performance has provided a better understanding of the differentiated impact technology has on the growth of individual industries and has highlighted important specificities of national systems of innovation. This is a field where the availability of improved – more significant and more comparable – databases will open new opportunities for research and policy studies.

6. INTELLECTUAL PROPERTY SYSTEMS AND THE APPROPRIATION OF BENEFITS FROM INNOVATIONS

A discussion of the nature of the intellectual property system – its institutional aspects, regulations, behaviour and policies – is beyond the scope of this chapter, but some issues need to be recalled here, as they affect firms' innovative activities and the significance of the indicators discussed so far, and raise important policy issues at the international level.

6.1. How firms appropriate the benefits of their innovations

The large body of literature on the appropriability of benefits from innovations has shown that the amount of resources devoted to innovation by firms is associated with the possibility of appropriating some of the returns. The relative importance of the various methods of appropriation have been assessed on the basis of survey responses given by managers and practitioners on the role of factors such as industrial secrecy, lead time, brand names, etc.

Besides the findings reported above (Section 2.2), the means of protecting innovations differ significantly among industries and countries. Wyatt *et al.* (1985) reported on the results of interviews with 80 multinationals in Europe, the United States and Japan. This study showed that patenting is considered to be much more effective by Japanese companies than by their American and European competitors. Levin *et al.* (1987) conducted a survey of more than 600 top US managers. This empirical evidence shows that the methods used to protect innovations differ substantially from industry to industry. These findings are broadly confirmed by the results of the EPO survey (EPO, 1994).

The PACE survey on technological appropriation recently completed for Europe shows that the highest shares of firms which use patent applications (firms applying for a patent for more than 80 per cent of their product innovations) are in pharmaceuticals (54 per cent of firms), followed by machinery, chemicals, telecommunications and automobiles (shares between 26 and 19 per cent). Conversely, in the basic metals sector, no firm makes extensive use of patents, and 77 per cent of firms apply for patents for from 0 to 19 per cent of their product innovations.

In addition, the shares of firms not applying for a patent in the previous three years ranges widely from 4 per cent in the plastics and pharmaceutical sectors to 31 per cent in the food industry. Across countries, the share of non-patenting firms ranges from 10 per cent in Germany to 15 per cent in Italy and in the smaller European countries, and to 17 per cent in the United Kingdom (Arundel *et al.*, 1995, pp. 60-61).

Once the results of the same survey are made available for the United States and Japan, the extent to which specific indicators, such as patents, are representative of the much broader innovative activities carried out by firms will be much clearer.

6.2. Patents as a tool of appropriability

Not only do the technical possibility of patenting inventions and the effectiveness of patent protection vary substantially across industries, the patenting policies of firms active in the same field may also differ. Some may use patents systematically to protect all inventions produced; others may patent only specific inventions and rely on secrecy or rapid technological change for others. Furthermore, firms often use patents to block competitors rather than to introduce innovations; this indicates that patents provide information not only on the innovative activity of firms, but also on their competitive strategies.

Empirical research has shown that in chemicals, chemical-related fields, and mechanical engineering, the propensity to patent is high. A large share of product innovations is patented, while process innovations are more likely to be protected by industrial secrecy.

There is a strong difference in the propensity to patent between market (profit-seeking) institutions and non-market (non-profit) ones, such as universities and government laboratories. In most countries, non-market institutions have less interest in protecting their scientific and technological output by patents, and their research is more likely to lead to the publication of scientific papers than to patent applications.

6.3. The objectives of national intellectual property systems

In moving from the microeconomic analysis of the appropriability strategies of individual firms to macroeconomic aspects and consequences, attention should be paid to the objectives and nature of national intellectual property systems. These have emerged from the need to find a balance between the different and sometimes contrasting interests that are relevant to an intellectual property system. Its basic requirements include:

- a balance between providing adequate incentives to generate invention and innovation (through strong protection of intellectual property), on the one hand, and ensuring rapid diffusion of new technologies, on the other;
- a balance between the private interest of the inventor, on the one hand, and society's interest in creating a stable system conducive both to invention and to diffusion, with due attention to welfare considerations, on the other;
- a balance between the temporary monopoly power granted to inventors by patent protection, on the one hand, and the norms for the protection of competition (such as anti-trust laws, industry regulations, etc.), on the other.

Within individual countries, a complex institutional and normative system has emerged, which balances the rights granted to innovators with measures for assuring the presence of competitive markets as well as efficiency and welfare considerations. The trade-offs between these different aspects, and in particular between the strength of the protection granted by the intellectual property system and the pace of diffusion of innovations, have been investigated in a large body of literature, which may take either an analytical or a policy approach, and which presents contrasting views and results.[6]

The links in this system are complex and multi-faceted. National and institutional aspects, the pattern of interaction among firms, the nature of the industrial structure, and behavioural aspects are all important in explaining how a country's innovation system operates. Furthermore, the outcome of a particular patent system depends on interaction with several other industrial regulations, such as licensing norms, anti-trust rules, and other forms of public policy (Ordover, 1991; Winter, 1993).

The innovative activity of firms has to be understood in the context of the specific intellectual property system in which they operate, but current changes in the innovation process, such as globalisation of firms' activity, greater co-operation, etc. (see OECD, 1992*b*), suggest the need to reconsider some aspects of the traditional intellectual property system.

Current changes in the innovation process, according to Foray (1994), result in greater emphasis on recombining and diffusing information, as exemplified by innovation by Japanese firms. In this perspective, a patent system designed to "reward" the inventor firm, by emphasizing its exclusive rights and monopoly power, is less effective than one where its role is to regulate the disclosure of information on research and innovation and to facilitate the diffusion of innovations. It has been shown that competing firms have a common interest in a freer exchange of information on the direction and success of their innovative efforts.

6.4. An international order for intellectual property systems

The problem of finding a balance between the different interests involved in an intellectual property system – incentives to invention vs. rapid diffusion, private vs. public interests, monopoly power vs. protection of competition – becomes all the more complex when set in an international context. Since the intensification of competition in the 1980s, in particular, the strategies of multinational firms and governments have given greater attention to intellectual property issues. The protection of innovations and technological know-how has been used as a competitive tool and has been the object of difficult trade disputes both among OECD countries and between developed and developing countries.

Over the last decades, divisions between the groups of advanced and developing countries have become stronger. Advanced countries have increased their economic and technological integration, thanks to increased exchange and similar market conditions and institutions. Two-way flows of technology among countries and firms with different specialisations are increasingly common, in a market context characterised by strong competition and regulation. Developing countries, instead, have continued to depend on foreign technology. It has been noted that "over 80 per cent of patents in third world countries are owned by foreigners, mainly TNCs [transnational corporations] of the United States, the United Kingdom, [the] FRG [the Federal Republic of Germany], France and Switzerland, and over 95 per cent are not used at all in production in these countries" (Raghaven, 1990, p. 123).[7] The issues of North-South technology flows have been largely ignored by recent innovation research and deserve more attention. Research should also take into account the emergence of newly industrialising countries (NICs) and industrialisation in several countries of the South, especially in Asia. In particular, the extent to which findings based on work limited to advanced countries holds for developing ones should be tested.

In the North-South context, the norms for intellectual property protection have been the object of difficult negotiations in the GATT Uruguay Round, as the United States, the European Union and Japan have sought to increase the protection granted to intellectual property though the inclusion in GATT of "Trade-Related Intellectual Property Rights" (TRIPs, which also include trade in counterfeit goods) (GATT, 1994; Otten, 1994). Although the issue of trade in this area has become more relevant, developing countries have questioned whether GATT is an institution for setting rules for the evolution of intellectual property systems (see Raghavan, 1990); they argue that the World Intellectual Property Organization (WIPO) or other UN organisations such as UNCTAD, UNESCO and UNIDO, could provide a more appropriate framework for discussion. Furthermore, attempts by advanced countries to enforce greater protection of intellectual property rights may work, to some extent, against a more liberalised and competitive trade system, by setting up excessive barriers to the entry of new producers and competitors.

At the international level and particularly in a North-South context, the balance between the different interests that an intellectual property system needs to reconcile can hardly be the same as the one that emerged in the specific context of advanced countries:

- the benefits granted to innovators largely flow to advanced countries, while the benefits from diffusion spread within developing countries;
- the private interest is that of multinational firms of advanced countries, while the public interest is that of developing countries;
- the counterbalances to the monopoly power granted to the inventor (protection of competition, anti-trust norms, etc.) are often missing in developing countries.

As industrialisation continues to spread, the definition of a multilateral intellectual property system, developed by consensus in an appropriate institutional setting, which takes into account the interests and needs of all countries and actors, may become increasingly necessary.

7. CONCLUSIONS

Patents and innovation surveys offer two important means of acquiring information about technological change in firms. They have advantages and disadvantages that complement other widely used indicators such as R&D, trade in high-technology products, bibliometric indicators, etc., and provide in-depth information on certain aspects which cannot be obtained from other indicators.

7.1. New developments for patent-based indicators

The use of patenting as an indicator of technological innovation has grown steadily over the past decade. Patent data are available at low cost and in large numbers; this incites international organisations, government agencies, research centres, and individual scholars to make use of them. Several of their shortcomings can now be dealt with either by statistical elaborations or by refinement of the data. The most relevant research areas and problems are:

- The need for improved estimates, more reliable and more comparable across countries, of the extent to which patents account for all inventions and are actually used for introducing innovations or carrying out production at the firm, industry and country levels.
- Refinement of the quality of patent data in order to understand the value of individual patents. As mentioned above (Section 2.2), patent citations, renewal fees, number of claims and number of extensions are used to account for the individual value of each patent. Over the last decade, the cost and effort required by this type of research have been considerably reduced by the availability of computerised databases. So far, individual research teams have invested their own resources and efforts to produce these data sources. *Is it desirable and feasible to obtain data on the quality of individual patents at the source, i.e. from the national patent offices? Could the efforts of the various teams converge in order to make this information more widely accessible?*
- Patents are also used to study technological interdependence. At present, only the Canadian patent office provides information on the sector of use of the patented invention. This information enlarges considerably the usefulness of patent indicators, and Canadian patents have been used by a growing number of scholars. It is to be hoped that other patent offices will follow this example. *What action should be taken in order to enrich standard patent statistics by including the main industry of use of the invention? Could the most important patent offices collect this information on a regular basis?*
- As indicated above (Table 2), one of the main advantages of patents is to provide internationally comparable information on detailed technological fields. Recent research has further improved the accuracy of patent-based international comparisons by examining the extension of patents to the most significant markets (see Schmoch and Kirsch, 1993). *How should this methodology be validated and developed further?*
- Patents are also increasingly used to monitor innovations occurring within the firm. Large firms have considerable innovative activities that are confidential in nature, but detailed descriptions of some of the most important innovative activities are reported in patent documents. *To what extent can patents be used to chart the direction and content of current innovative activities carried out by firms?*
- Finally, a large body of literature has used patent-based indicators either at country or at industry level in order to link technology to patterns in science, R&D, production and exports. *Is it possible to develop a more systematic*

approach to relating patents to other data, including technology, bibliometric and economic indicators?

7.2. New developments for innovation surveys

Innovation surveys are one of the main new developments in the measurement of technological change. Until very recently, they faced some basic limitations, notably the lack of harmonisation and standardisation over time and across countries. The OECD Oslo Manual and the EU-sponsored Community Innovation Survey (CIS) have made available a large body of evidence, with internationally comparable and standardised data, for the countries of the European Union. The main research areas and problems in this field are:

- Besides the core questions of the OECD-EU harmonised questionnaire used in the CIS-based innovation surveys, the surveys could gather information on a wide variety of themes of potential interest to researchers and policy makers. Agreement on which aspects deserve priority is important, in order to make more effective the process of revising the Oslo Manual and preparing for the next round of the Community Innovation Survey.
- Innovation surveys carried out so far have indicated that, to obtain comparable results, a common questionnaire is a necessary but by no means sufficient condition. Survey results can be compared only if the statistical methodology used, including implementation and sampling, is harmonised. Efforts should be made to establish a common statistical methodology for the basic analysis of innovation data.
- Innovation surveys do not yet provide for carrying out time-series comparisons. It is now feasible and useful to start collecting data on innovation at regular intervals; it has been recommended that innovation surveys should be carried out every three years.
- Out of the large amount of information emerging from the CIS surveys, it would be important to identify new key variables that summarise effectively the evidence provided by the survey. In particular, variables such as the total cost of innovation, the share of sales due to innovative products, and the specific role of investment emerge as new major indicators derived from the CIS surveys. For each, it is necessary to validate their relevance and compare the evidence they provide with that p rovided by other indicators.
- Besides the specific quantitative indicators obtained, innovation surveys offer an important opportunity to investigate the deeper mechanisms of the innovation process, including the objectives pursued, the sources used, the obstacles encountered. In particular, evidence on the objective of innovative efforts may shed new light on the process by which firms select innovations,

on their quality and nature, and on their relation to organisational changes and to major current problems, such as the diffusion of information technology, technological unemployment, and environmental protection.

Finally, the results of the parallel developments in the fields of patent and innovation indicators should be compared in order to provide a more integrated picture of innovative activities at the firm and industry level. This would make it possible to give a more adequate quantitative description of the current technological strategies in OECD countries and thus provide support for policy decisions of firms and governments.

7.3. Key issues for further research

The field of innovation indicators requires close interaction between concepts and measures, between method and theory. It is expanding rapidly, and new developments are emerging in the context of different studies and disciplines. Researchers in various disciplines are increasingly using technology indicators in different conceptual contexts; they range from technologists and business specialists to economists (who study firms, industries or the macroeconomy), and from legal experts on the appropriability of inventions to policy makers in science, technology and industry. They all make different assumptions and ask different questions. While the wide use of technology indicators should be encouraged, it is important to ensure interaction among the different disciplines, so as to stimulate continuing discussion and progress in the basic understanding of how technology indicators relate to changes in science, technology, economy and society.

The OECD workshop and the studies included in this volume point to a set of crucial themes, including both methodological problems and areas of further research.

The need for a common framework and greater comparability

It is often difficult to relate new findings to the previous body of evidence, as the specific design and methodology of individual studies (especially in the case of innovation surveys) may result in a serious lack of comparability. Therefore, a strong frame of reference is needed, so that the findings of individual studies, based on specific indicators, can be located in the broader landscape of the evidence on innovative activities provided by the multiple indicators available.

Further, as there is already a very large body of research on patent data and as information from comparable innovation surveys is becoming available, it will be possible to achieve a more comprehensive and systematic understanding of the relationships between these indicators and the more established data on R&D expenditure. The interpretation of results of studies using different technology indicators would benefit from the availability of a "benchmark" reference tool, which would provide a comparative framework for assessing the nature, pattern and impact of innovative activities. In particular, it is important to show more precisely the degree of overlap, consistency and complementarity of the different indicators.

The availability of the new CIS-based innovations surveys in Europe is expected to generate many studies. It is important to make a systematic effort to relate the new findings to the evidence already provided by other indicators. Important issues in this regard include the role of R&D in the context of total innovation expenditure, the role of innovation-related investment, and the market impact of innovation as measured by the share of innovative sales.

The "next" innovation indicators: services, software, usage and objectives of innovations

The range of innovative activities examined by technology indicators needs to be expanded. While most efforts so far have been confined to the manufacturing industry, the CIS survey is now being tested for the inclusion of service industries, which are major users of innovations, namely advanced information technology. Here, the very concept of technological innovation has to be clarified, and progress must also be made on the criteria for the classification of service activities. Software, which represents a major area of innovation across manufacturing and service industries, presents a particular problem, since it is covered by copyright rather than by patent protection.

Interest in the nature and relevance of organisational innovations within firms and in business networks is increasing. Important research questions include how to define and quantify them and how they relate to technological innovations.

So far, most emphasis has been on the production of innovations; however, more data is now becoming available on the use to which they are put, particularly in the case of information technology goods and services. Much can be learned about the role of innovation in economic change by tracing, with appropriate indicators, the introduction and diffusion of innovations among users.

More attention should also be devoted to the information that can be gathered on the objectives and priorities of innovative efforts, and, in particular, on their impact on employment, the environment and society.

Appropriability of innovations and the role of patents

Much attention is currently being devoted to the increasing codification of knowledge and to its relevance for industrial innovation. Recent results from European surveys have confirmed that patents, a classical case of codified knowledge, are a very good indicator of firms' innovative activity. However, areas where additional research is needed include the relationship between codified and tacit knowledge in firms and industries, the flows of know-how in national innovation systems and within multinational companies, the impact of patents on the diffusion of innovations, and the efficiency of national intellectual property systems.

The classification and concordance of patent data so that they can be used in studies of firms' and sectoral performances are major and continuing problems. There is also scope for using specific patent indicators (*e.g.* citations, co-inventors, etc.) in highly focused studies of technology areas and of the interaction between science, technology and industrial activity.

The industrial structure of innovation: sectoral specificities and flows of technology

When technological indicators are aggregated at the sectoral level, two major issues arise. First, sectoral specificities of innovative activities need to be pointed out, considering the sectoral bias of individual indicators. Here again, a clearer comparative framework for assessing the results of different indicators would be useful.

Second, more attention should be paid to the flows of innovations from the industries of origin to the industries of use. This issue can be addressed by using evidence from innovation surveys and the Canadian patent matrix, as well as input-output matrixes for the flows of intermediate and investment goods.

Innovation and performance

A growing literature has investigated the impact innovative activities have on the performance of firms, sectors and countries. Once again, a common framework would be useful, especially for addressing questions such as the comparative patterns of small and large firms, the dynamics of rapidly growing sectors, the distribution of the benefits of technological change from the firms and industries producing the innovations to those using them.

However, the complexity of the relationship between the innovation process and performance should not be overlooked; without an adequate understanding of the causal relationships and of the dynamics of innovation, there is the risk that the innovative activities of firms will be "explained" by their degree of "success" and that their successful performance will be "explained" by their greater innovative efforts.

Innovation and the changing boundaries of the firm

Innovation is a heterogeneous phenomenon; in addition, the structure of firms that introduce innovations has become increasingly differentiated. While it is often assumed that the entire innovative process is carried out within a single firm, innovating firms draw in fact from a much broader range of knowledge and activities.

The boundaries of innovating firms are in fact changing. The importance of large in-house R&D laboratories is declining, while inter-firm R&D co-operation, the role of small innovating firms affiliated to larger companies, the network structure of innovators, the closer links to university research, and the growing international dimension of the innovation process are increasing. More generally, the rapid pace of organisational innovations within business groups is also reshaping the scope and nature of innovation within firms.

Globalisation of technology and national systems of innovation

The increasingly transnational nature of innovative activities raises important problems for technology indicators. The generation, introduction and appropriation of benefits from innovation take place less and less within a single country, most obviously for countries of smaller size and stronger international orientation. The growing role of multinational firms suggests that the study of innovation and its economic impact increasingly requires a scale of analysis that transcends national economies.

How this phenomenon relates to the characteristics of national systems of innovation is a major question for both measuring and conceptualising innovative activities and for policy.

Policy issues

Innovation and technology are of major importance to policy. The increasing opening up of national markets and the reduction of government intervention in industry mainly leave room for public policies that support the technological capabilities and knowledge base of countries.

Furthermore, many essential aspects of technology and innovation are affected or regulated by governments (through patent regulations, the setting of standards, public procurement, anti-trust rules, etc.); there is a broad scope for public policy in the definition of the tools, priorities and objectives of the innovation process in the national economy.

Innovation-related policy issues are also relevant, however, at the supranational level. The rapid globalisation of technology means that the rules for cross-border flows of technology need to be updated, and the inclusion in the GATT agreement of issues of intellectual property has gone in this direction. However, much more is needed to develop an appropriate international setting for technology.

Finally, while many of the issues raised in this section have originated from empirical problems of research on technology indicators, they can only be successfully addressed by improving the theoretical understanding of how technological change takes place in modern economies and societies. Again, close association and interaction between advances in concepts and in empirical work, between methodology and policy, appear to be indispensable.

NOTES

1. Patent-based empirical research is reviewed in Basberg, 1987; Pavitt, 1988; and Griliches, 1990.
2. Recent research has shown that the average number of claims per patent has considerably increased over the last 20 years and that significant differences are found across countries (Tong and Frame, 1994).
3. The object approach has also been experimented at the country level (see the study of significant Swedish innovations by Wallmark and McQueen, 1991) and for selected technological areas (see the study on the pesticide industry by Achilladelis *et al.*, 1987).
4. Achilladelis *et al.* (1987) and Acs and Audretsch (1989) have compared patents to innovation counts.
5. This extensive literature is reviewed, among others, by Kamien and Schwartz (1982) and Cohen and Levin (1989).
6. While firms have traditionally argued for greater patent protection, one study has concluded that "very short-lived patents, or even no patents at all, might in some contexts yield higher economic welfare than the patent rights conferred under existing institutional arrangements" (Winter, 1993, p. 211).
7. No comparison of data for OECD countries is possible. As noted above, studies such as Scherer *et al.* (1959) and Napolitano and Sirilli (1990) have shown that the share of patents actually used by the inventing firms ranges between 40 and 60 per cent, but no account is taken of patents filed by foreign firms.

REFERENCES

ACHILLADELIS, B., A. SCHWARZKOPF and M. CINES (1987), "A Study of Innovation in the Pesticide Industry: Analysis of the Innovation Record of an Industrial Sector", in C. Freeman (ed.), *Output Measurement in Science and Technology*, pp. 175-202, North-Holland, Amsterdam.

ACS, Z.J. and D.B. Audretsch (1989), "Patents as a Measure of Innovative Activity", *Kyklos*, Vol. 42, No. 2, pp. 171-180.

ACS, Z.J. and D.B. AUDRETSCH (1990), *Innovation and Small Firms*, MIT Press, Cambridge, Massachusetts.

ACS, Z.J. and D.B. AUDRETSCH (1991), *Innovation and Technological Change: An International Comparison*, Harvester Wheatsheaf, New York.

ACS, Z.J. and D.B. AUDRETSCH (1992), "Technological Regimes, Learning, and Industry Turbulence", in F.M. Scherer and M. Perlman (eds.), *Entrepreneurship, Technological Innovation, and Economic Growth*, pp. 305-320, University of Michigan Press, Ann Arbor, Michigan.

AMENDOLA, G., G. DOSI and E. PAPAGNI (1993), "The Dynamics of International Competitiveness", *Weltwirtschaftliches Archives*, Vol. 129, No. 3, pp. 451-471.

AMENDOLA, G., P. GUERRIERI and P.C. PADOAN (1992), "International Patterns of Technological Accumulation and Trade ", *Journal of International and Comparative Economics*, Vol. 1, No. 1, pp. 173-197.

ARCHIBUGI, D. (1988), "In Search of a Useful Measure of Technological Innovation, *Technological Forecasting and Social Change*, Vol. 34, No. 3, pp. 253-277.

ARCHIBUGI, D., S. CESARATTO, G. SIRILLI (1991), "Sources of Innovative Activities and Industrial Organization in Italy", *Research Policy*, Vol. 20., pp. 299-313.

ARCHIBUGI, D. and M. PIANTA (1992), *The Technological Specialization of Advanced Countries*, Kluwer, Dordrecht.

ARCHIBUGI, D., P. COHENDET, A. KRISTENSEN and K.-A. SCHAFFER (1995), *Evaluation of the Community Innovation Survey*, IKE Group, Aalborg-Sprint/Eims Report.

ARCHIBUGI, D., R. EVANGELISTA, R. SIMONETTI (1995), "Concentration, Firm Size and Innovation. Evidence from Innovation Costs", *Technovation*, Vol. 15, No. 3, pp. 153-163.

ARCHIBUGI, D. and J. MICHIE (1995), "The Globalisation of Technology. A New Taxonomy", *Cambridge Journal of Economics*, Vol. 19, No. 1, pp. 121-140.

ARUNDEL, A. G. van de PAAL and L. SOETE (1995), *Innovation Strategies of Europe's Largest Industrial Firms: Results of the PACE Survey on Information Sources, Public Research, Protection of Innovation and Government Programmes*, MERIT, Maastricht.

BARRÉ, R. *et al.* (1994), *Science et Technologie Indicateurs*, Economica, Paris.

BASBERG, B.L. (1987), "Patents and the Measurement of Technological Change: A Survey of the Literature, *Research Policy*, Vol. 16, Nos. 2-4, pp. 131-141.

CANTWELL, J. (1989), *Technological Innovation and Multinational Corporations*, Basil Blackwell, Oxford.

CASSON, M. (ed.) (1991), *Global Research Strategy and International Competitiveness*, Basil Blackwell, Oxford.

CESARATTO, S. and S. MANGANO (1992), "Technological Profiles and Economic Performance in the Italian Manufacturing Sector", *Economics of Innovation and New Technology*, Vol. 2, pp. 237-256.

COHEN, W.M. and R.C. LEVIN (1989), "Empirical Studies of Innovation and Market Structure", in R. Schmalensee and R.D. Willig (eds.), *Handbook of Industrial Organization*, Vol. 2, pp. 1059-1107, Elsevier, Amsterdam.

CREPON, B. and E. DUGUET (1994), "Innovation: Mesures, rendements et concurrence", *Economie et statistique*, Nos. 275-276, pp. 121-134.

DeBRESSON, C. (1990), "L'analyse inter-industrielle et le changement technologique", *Revue d'économie politique*, Vol. 100, No. 6, pp. 833-869.

DeBRESSON, C., G. SIRILLI, X. HU, F.K. LUK (1994), "Structure and Location of Innovative Activity in the Italian Economy, 1981-85", *Economic Systems Research*, Vol. 6, No. 2, pp. 135-158.

DOSI, G., K. PAVITT and L. SOETE (1990), *The Economics of Innovation and International Trade*, Harvester Wheatsheaf, New York.

DUGUET, E. (1994), "La coopération technique au travers des co-brevets européens", *Economie et statistique*, No. 275-276, pp. 135-148.

ENGLANDER, A.S., R. EVENSON and M. HANAZAKI (1988), "R&D, Innovation, and the Total Factor Productivity Slow Down, *OECD Economic Studies*, Vol. 11, pp. 8-42.

EUROPEAN PATENT OFFICE (1994), *Utilisation of Patent Protection in Europe*, European Patent Office, Munich.

EUROPEAN PATENT OFFICE, JAPANESE PATENT OFFICE and US PATENT AND TRADEMARK OFFICE (1993), *Trilateral Statistical Report 1993*, EPO, JPO, USPTO, Munich.

EUROSTAT (1994), *The Community Innovation Surveys. Status and Perspectives*, Eurostat, Luxembourg.

FAGERBERG, J. (1987), "A Technology Gap Approach to Why Growth Rates Differ" in C. Freeman (ed.), *Output Measurement in Science and Technology*, pp. 87-100, North-Holland, Amsterdam.

FAGERBERG, J. (1988), "International Competitiveness", *Economic Journal*, Vol. 98, pp. 355-374.

FORAY, D. (1994), "Production and Distribution of Knowledge in the New Systems of Innovation: The Role of Intellectual Property Rights", *STI Review*, No. 14, pp. 119-152.

FREEMAN, C. (1992), *The Economics of Hope,* Pinter, London.

GATT (1994), *Agreement on Trade-related Aspects of Intellectual Property Rights*, Annex 1C, GATT, Geneva.

GRILICHES, Z. (1990), "Patent Statistics as Economic Indicators: A Survey", *Journal of Economic Literature*, Vol. 28, No. 4, pp. 1661-1797.

GRILICHES, Z. (ed.) (1984), *R&D, Patents and Productivity*, University of Chicago Press, Chicago.

GRUPP, H. (1991), "Innovation Dynamics in OECD Countries: Towards a Correlated Network of R&D-intensity, Trade, Patent and Technometric Indicators, pp. 275-295, in OECD (ed.), *Technology and Productivity: The Challenge for Economic Policy*, Paris.

GRUPP, H. (ed.) (1992), *Dynamics of Science-based Innovation*, Springer-Verlag, Berlin.

GRUPP, H. (1993), "Dynamics of Science-based Innovation in Northern America, Japan and Western Europe", in S. Okamura, F. Sakauci and I. Nonaka (eds.) (1993), *Science and Technology Policy Research: New Perspectives on Global Science and Technology Policy*, Mita Press, Tokyo.

HAGEDOORN, J. and J. SCHANKERAAD (1993), "Strategic Technology Partnering and International Corporate Strategies", in K. Hughes (ed.), *European Competitiveness*, pp. 60-86, Cambridge University Press, Cambridge.

HANEL, P. (1993), "Interindustry Flows of Technology: An Analysis of the Canadian Patent Matrix and Input-output Matrix for 1979-1989", paper presented at the International Input-Output Conference, Seville.

HANSEN, J.A. (1992), *New Indicators of Industrial Innovation in Six Countries: A Comparative Analysis*, Final Report to the US National Science Foundation, June.

von HIPPEL, E. (1988) *The Sources of Innovation*, Oxford University Press, New York.

ISTAT (1989), *Indagine statistica sull'innovazione tecnologica nell'industria italiana. Anni 1981-85*, ISTAT, Rome.

ISTAT (1995), "Indagine sull'innovazione tecnologica, anni 1990-92", *Notiziario ISTAT*, Vol. 16, No. 2, June.

JAFFE, A. (1986), "Technological Opportunity and Spillovers of R&D: Evidence from Firms' Patents, Profits and Market Value", *American Economic Review*, Vol. 76, No. 5, pp. 984-1001.

KABLA, I. (1994), "Un indicateur de l'innovation : le brevet", *Economie et statistique*, Nos. 275-276, pp. 95-110.

KAMIEN, M.I. and N.L. SCHWARTZ (1982), *Market Structure and Innovation,* Cambridge University Press, Cambridge.

KAMINSKI, P. (1993), "Comparison of Innovation Survey Findings", paper for the EC-OECD Joint Seminar on Innovation Surveys, Paris, 29 April.

KLEINKNECHT, A. (1987), Measuring R&D in Small Firms: How Much Are We Missing?, *The Journal of Industrial Economics*, Vol. 36, No. 2, pp. 253-256.

KLEINKNECHT, A. and J.O.N. REIJNEN (1991), "More Evidence on the Undercounting of Small Firm R&D, *Research Policy,* Vol. 20, pp. 579-592.

KLEINKNECHT, A. and D. BAIN (eds.) (1993), *New Concepts in Innovation Output Measurement,* Macmillan, Houndmill.

KRISTENSEN, A. (1993), "Innovation Structures and Performance in Nordic Manufacturing Industry", IKE Group, University of Aalborg, May.

KODAMA, F. (1986), "Technology Diversification of Japanese Industry", *Science*, Vol. 233, pp. 291-233.

LEVIN, R.C., A.K. KLEVORIK, R.R. NELSON and S. WINTER (1987), "Appropriating the Returns From Industrial Research and Development", *Brookings Papers on Economic Activity, Microeconomics*, No. 3.

LICHT, G. (1994), "Patenting, Innovation and R&D: Results from the German CIS", paper for the OECD workshop on *Innovation, Patents and Technological Strategies*, Paris, 8-9 December.

MALERBA, F. and L. ORSENIGO (1995), "Schumpeterian Patterns of Innovation", *Cambridge Journal of Economics*, Vol. 19, No. 1, pp. 47-65.

MANSFIELD, E. (1986), "Patents and Innovation: An Empirical Study", *Management Science*, Vol. 32, No. 2, pp. 173-81.

MINISTÈRE DE L'INDUSTRIE (1994), *L'innovation technologique dans l'industrie*, SESSI, Paris.

NAPOLITANO, G. and G. SIRILLI (1990), "The Patent System and the Exploitation of Inventions: Results of a Statistical Survey Conducted in Italy", *Technovation*, Vol. 10, No. 1, pp. 5-16.

NARIN, F. and D. OLIVASTRO (1988), "Technology Indicators Based on Patents and Patent Citations" in A.F.J. Van Raan (ed.), *Handbook of Quantitative Studies of Science and Technology*, pp. 485-506, Elsevier, Amsterdam.

NIWA, F. (1992), "The Measurement of Technology Diversification Using Patent Applications", in *Science and Technology Policy Research,* pp. 123-138, Mita Press, Tokyo.

OECD (1992*a*), "OECD Proposed Guidelines for Collecting and Interpreting Technological Innovation Data – Oslo Manual", OCDE/GD(92)26, Paris.

OECD (1992*b*), *Technology and the Economy: The Key Relationships*, Paris.

OECD (1994), "The Measurement of Scientific and Technological Activities: Using Patent Data as Science and Technology Indicators – Patent Manual", OCDE/GD(94)114, Paris.

ORDOVER, J.A. (1991), "A Patent System for Both Diffusion and Exclusion", *Journal of Economic Perspectives*, Vol. 5, No. 1, pp. 43-60.

OTTEN, A. (1994), "Trade-related Intellectual Property Rights: What Will Change?, paper presented at the International Conference on Intellectual and Industrial Property: Objectives and Strategies, 8 April.

PAKES, A. and M. SIMPSON (1989), "Patent Renewal Data", *Brookings Papers of Economic Activity, Microeconomics*, pp. 331-410.

PATEL, P. (1995), "Localised Production of Technology for Global Markets", *Cambridge Journal of Economics,* Vol. 19, No. 1, pp. 141-154.

PATEL, P. and K. PAVITT (1987), "Is Western Europe Losing the Technological Race?", *Research Policy*, Vol. 16, pp. 59-85.

PATEL, P. and K. PAVITT (1991), "Large Firms in the Production of World's Technology: An Important Case of Non-globalisation", *Journal of International Business Studies,* No. 1, pp. 1-21.

PATEL, P. and K. PAVITT (1994), *Technological Competences in the World's Largest Firms,* Science Policy Research Unit, Brighton.

PAVITT, K. (1984), "Sectoral Patterns of Technological Change: Toward a Taxonomy and a Theory, *Research Policy,* Vol. 13, pp. 343-373.

PAVITT, K. (1988), "Uses and Abuses of Patent Statistics" in A.F.J. Van Raan (ed.), *Handbook of Quantitative Studies of Science and Technology,* pp. 509-536, Elsevier, Amsterdam.

PAVITT, K., M. ROBSON, J. TOWNSEND (1987), "The Size Distribution of Innovating Firms in the UK: 1945-1983", *The Journal of Industrial Economics,* Vol. 35, No. 3, pp. 297-316.

PEARCE, R. and S. SINGH (1992), *Globalizing Research and Development,* Macmillan, Houndmills.

RAGHAVAN, C. (1990), *Recolonization. GATT, the Uruguay Round & the Third World,* Zed Books, London.

ROBSON, M., J. TOWNSEND and K. PAVITT (1988), "Sectoral Patterns of Production and Use of Innovations in the UK: 1945-1983", *Research Policy,* Vol. 17, pp. 1-14.

SCHERER, F.M. (1982*a*), "Inter-industry Technology Flows and Productivity Growth", *Review of Economics and Statistics,* Vol. 64, pp. 627-634.

SCHERER, F.M. (1982*b*), "Inter-industry Technology Flows in the US", *Research Policy,* Vol. 11, No. 4, pp. 227-245.

SCHERER, F.M. *et al.* (1959), "Patents and the Corporation", Boston, mimeo.

SCHMOCH, U. and N. KIRSCH (1993), "Analysis of International Patent Flows. Final Report to the OECD", FhG-ISI, Karlsruhe.

SCHMOOKLER, J. (1966), *Invention and Economic Growth,* Harvard University Press, Cambridge, Massachusetts.

SCHWITALLA, B. and H. GRUPP (1994), "Measuring and Explaining Industrial Innovation Activities. An Empirical Analysis of West German Firms", paper presented at the Conference on Productivity, R&D and Innovation at the Firm Level, Mannheim, June 22-24.

SIMONETTI, R. (1994), "The Dynamics of Market Shares: A Disequilibrium Model", paper presented at the EUNETICS Conference, Strasbourg, October.

SIRILLI, G. (1987), "Patents and Inventors: An Empirical Study" in C. Freeman (ed.), *Output Measurement in Science and Technology,* pp. 157-172, North-Holland, Amsterdam.

SMITH, K. (1992*a*), "Technological Innovation Indicators: Experience and Prospects", *Science and Public Policy,* Vol. 19, No. 6, pp. 383-392.

SMITH, K. (1992*b*), *Quantitative Innovation Studies in Europe with Existing Datasets: Possibilities and Problems,* Royal Norwegian Council for Scientific and Industrial Resarch, Oslo.

SOETE, L. (1981), "A General Test of the Technological Gap Trade Theory", *Weltwirtschaftliches Archiv,* No. 4, pp. 638-659.

SOETE, L. (1987), "The Impact of Technological Innovation on International Trade Patterns: The Evidence Reconsidered" in C. Freeman (ed.), *Output Measurement in Science and Technology*, pp. 101-137, North-Holland, Amsterdam.

STERLACCHINI, A. (1989), "R&D, Innovations, and Total Factor Productivity Growth in British Manufacturing, *Applied Economics*, Vol. 21, pp. 1549-1562.

SWANN, P. (1993), "Inference from Mixed Bags: The Economic Value of Patent Counts, Innovation Counts and the Like", London Business School, mimeo.

TONG, X. and J.D. FRAME (1994), "Measuring National Technological Performance with Patent Claims Data, *Research Policy*, Vol. 23, pp. 133-141.

TOWNSEND, J., F. HENWOOD, G. THOMAS, K. PAVITT and S. WYATT (1981), "Innovations in Britain Since 1945", Occasional paper No. 16, SPRU, University of Sussex, Falmer.

TRAJTENBERG, M. (1990), *Economic Analysis of Product Innovation: The Case of CT Scanners*, Harvard University Press, Cambridge, Massachusetts.

Van RAAN, A.F.J., A J NEDERHOF, A.J. and H F MOED (eds.) (1989), *Science and Technology Indicators: Their Use in Science Policy and Their Role in Science Studies*, DSWO Press, Leiden.

VERSPAGEN, B. (1993), *Uneven Growth Between Interdependent Economies*, Avebury, Aldershot.

VERSPAGEN, B., T. von MOERGASTEL, and M. SLABBERS (1994), *Merit Concordance Table: IPC-ISIC (Rev. 2)*, MERIT, Maastricht.

VIVARELLI, M., R. EVANGELISTA and M. PIANTA (1995), "Innovation and Employment: Evidence from Italian Manufacturing", paper presented at the Conference on The Effects of Technology and Innovation on Firm Performance and Employment, Washington DC, 30 April-2 May.

WALLMARK, J.T. and D.H. McQUEEN (1991), "One Hundred Major Swedish Technical Innovations, from 1945 to 1980", *Research Policy*, Vol. 20, pp. 325-344.

WINTER, S. (1993), "Patents and Welfare in an Evolutionary Model", *Industrial and Corporate Change*, Vol. 2, No. 2, pp. 209-223.

WYATT, S., G. BERTIN, G. and K. PAVITT (1985), "Patents and Multinational Corporations: Results from Questionnaires, *World Patent Information*, Vol. 7, pp. 196-212.

A COMPARISON OF R&D SURVEYS, INNOVATION SURVEYS AND PATENT STATISTICS BASED ON FINNISH DATA

by
Mikael Åkerblom
Eurostat, detached from Statistics Finland
and
Markku Virtaharju and Ari Leppälahti
Statistics Finland

INTRODUCTION

Most OECD countries have collected statistics on resources devoted to R&D since the 1960s. During the 1970s and 1980s, constantly increasing amounts of money were devoted to R&D, and interest in obtaining quantitative information in order to evaluate the results and effects of this expenditure has increased accordingly. To satisfy these needs, various R&D output indicators have been developed since the 1970s:

- bibliometric indicators (number of articles, papers, citations, co-citations);
- technological balance of payments (payments and receipts for technology);
- trade in high-technology products (mainly products with high R&D intensity);
- patent indicators (patent applications, patents granted);
- innovation indicators.

The last of these is a new statistical system that is being developed. It is based on the provisional guidelines of the OECD Oslo Manual; these have been used by Eurostat for the Community Innovation Survey (CIS). In co-operation with the OECD, Eurostat has developed a questionnaire, which has been used for innovation surveys in all EU member States. Corresponding surveys have also been made in most non-EU OECD Member countries. The innovation surveys include questions on sources of innovation, objectives of innovation, barriers to innovation, technology acquisition, R&D co-operation, innovation expenditure, sales and exports due to innovative products.

Efforts to compare the results derived from various indicators so as to give a consistent message to policy makers have been rather limited. The indicators most suitable for such comparisons are the patent and innovation indicators; like the traditional R&D inputs indicator, they can be analysed at the level of industries.

Internationally comparable data are available for R&D inputs at the industry level. For patent statistics, some international comparisons can be made using patents granted in the United States, classified according to a US scheme for classifying patents according to the standard national industrial classification, and patent applications to the European Patent Office (EPO), classified according to the industrial classification using a transformation key, developed by MERIT, for the international patent classification (IPC) and the Standard Industrial Classification. This analysis is included in the first *European Science and Technology Indicators Report*, which was published under the responsibility of DG XII of the European Commission in November 1994. In co-operation with Eurostat, EPO is developing a correspondence key between IPC and NACE (Statistical Classification of Economic Activities in the European Community) which will be used in the future for more extensive European comparisons.

Some comparable data on innovation will be available as soon as the data from the CIS are ready. For the moment, there are not enough international data available to compare the results of surveys of R&D input and of innovation and patent indicators. At national level, it is easier to make such comparisons, especially when the same institution is responsible for R&D surveys, innovation surveys and patent statistics at industry level, as is the case for Statistics Finland.

This chapter uses data from Statistics Finland to study certain indicators for R&D, innovations and patents at industry level in order to analyse the relationships among the indicators. It is an attempt to evaluate the usefulness of different indicators for analysing the relationships between R&D input and its results and effects.

DATA AND METHODOLOGY

Statistics Finland carries out surveys of R&D activities in the business enterprise sector every second year. The data are classified both according to firms' main activities and according to the orientation of R&D by product field. The information used here is for 1985-91.

Statistics Finland has also made two experimental surveys of technological innovation. The second of these, for 1991, the first according to the Oslo Manual and quite similar to the CIS, has been used here. The data have been classified according to the main activity of the firm.

Statistics Finland regularly produces statistics on patent applications filed at the Finnish National Patent and Register Board. In order to have a more complete picture, as well as for practical reasons, we have used national applications. To be able to compare the results with information from enterprise surveys, we have only included applications originating from firms. The information is drawn from the patent database maintained by the Finnish National Patent and Register Board. The data used are for 1985-92, and patent applications have been classified by industry according to the concordance key between the IPC and the Finnish Industrial Classification (available on request from Statistics Finland). The classification is determined by the assumed end use of the product or process of the patent applied for.

The surveys described above were used to compare aggregated results at industry or product group level. The results of the analysis are preliminary and will be further elaborated and published in the next science and technology indicator publication of Statistics Finland.

RESULTS

The following tables rank industries according to various intensities. All calculations are based on figures grossed up from sample values to total industrial level (see also Annex Table 1).

The most innovative industries are not always the most R&D-intensive

Rank correlation coefficients in Table 1 were calculated to obtain a general view of rankings. The high correlation between R&D and innovation intensities is of course self-evident, because R&D expenditures are included in total innovation costs. The correlation between ranking by the calculated intensities and ranking by sales due to innovative products is considerably lower.

There is a strong relation between R&D intensity and innovativeness for the most R&D-intensive and most innovative industry (electrotechnical industry) and the least R&D-intensive and least innovative industry (publishing and printing). A good match is also found for machinery, which ranks high according to both measures.

Chemical manufacturing has a much lower innovativeness ranking. In addition, owing to investments in innovative process equipment, pulp and paper ranks higher for innovativeness than for R&D intensity. The mismatch with the innovativeness measure for these industries is due to the fact that the outcome of innovation activities is typically a process innovation, a feature that is not covered by the innovativeness measure. Otherwise, the rankings are more or less the same for R&D intensities and innovativeness.

Table 1. **Industries ranked according to R&D intensity (R&D expenditure/sales), innovation intensity (total innovation expenditures including innovated-related investments) and innovativeness (sales due to innovative products), 1991**

	Rank according to R&D intensity	Rank according to innovation intensity	Rank according to sales due to innovative products
Electrotechnical	1	1	1
Basic metals	2	2	14
Machinery	3	6	3
Metal products	4	7	13
Chemical products	5	12	10
Other industrial products	6	4	8
Transport equipment	7	5	2
Pulp and paper	8	3	11
Stone, clay and glass	9	10	6
Textiles	10	14	5
Food	11	13	12
Clothes	12	11	4
Furniture	13	8	7
Wood and wood products	14	9	9
Publishing and printing	15	15	15
Spearman correlation coefficient between R&D intensity and innovation intensity and sales of innovative products		0.7	0.2

Source: Statistics Finland.

Furniture, textiles and clothes rank higher for the innovativeness measure than for the R&D measure. This might be explained by the importance of design in the innovation process and short life cycles. Metal products and basic metals are rather R&D-intensive but score considerably lower for innovativeness; for basic metals the reason probably is the strong emphasis on process-related R&D and process innovations, which are not covered by the innovativeness measure.

Innovation intensity is calculated by combining R&D intensities from the R&D survey and intensities for other innovation expenditure, as calculated from the innovation survey. It was not possible to use R&D expenditure as a part of innovation expenditure from the innovation survey, owing to difficulties in obtaining reliable data on R&D from the Finnish innovation survey. The same is, of course, true for other innovation expenditure, but, in the absence of a better alternative, these data have been used.

The correlation between R&D intensity and innovativeness variables and between innovation intensity and innovativeness variables at firm level was analysed, on the basis of firms responding to the Finnish innovation survey. No significant correlation was found, either at total industry level or for some key industries. There was a small intra-industry variance in the R&D intensity of firms but more varied innovativeness variables between firms. This resulted in non-significant correlation.

An indicator based on the share of highly innovative firms in Finnish industry gives a different picture

The rankings in Table 2 use an alternative measure of the innovativeness of an industry. A firm is considered highly innovative if it fulfills all of the following conditions:

- performed R&D in 1991;
- acquired and sold technology in 1991;
- introduced both product and process innovations in 1989-91.

The order of industries is maintained to facilitate comparisons.

According to these output measures, the rankings are rather different. Textile manufacturing has the highest share of highly innovative firms. Pulp and paper has the second highest share of highly innovative firms, considerably higher than the ranking according to sales due to innovative products.

Electrical products and machinery rank lower than for sales due to innovative products. Basic metals, chemical products, publishing and printing also rank higher according to this measure. The differences can again be explained by process innovations, which are more explicitly covered by the first indicator.

Table 2. **Industries ranked according to the share of highly innovative firms in the industry and innovativeness in terms of sales due to innovative products, 1991**

	Rank according to share of highly innovative firms	Rank according to sales due to innovative products
Electrotechnical	4	1
Basic metals	6	14
Machinery	10	3
Metal products	13	13
Chemical products	3	10
Other industrial products	15	8
Transport equipment	9	2
Pulp and paper	2	11
Stone, clay and glass	5	6
Textiles	1	5
Food	11	12
Clothes	8	4
Furniture	14	7
Wood and wood products	12	9
Publishing and printing	7	15

Source: Statistics Finland.

R&D: a more important aspect of the innovation process than assumed

The following simple comparisons between firms performing R&D and firms that have introduced innovations are based on the unweighted material in the innovation survey. Of the 876 firms responding, 598, or 68 per cent, introduced a product or process innovation in 1989-91. Of the firms introducing innovations, 487 firms introduced an incremental product innovation, 432 introduced a process innovation, and 419 introduced a major product innovation. Of the 598 firms that have introduced innovations, 550, or 92 per cent, reported R&D during 1991. Of the 263 firms not reporting any R&D in 1991, only 48, or 18 per cent, reported the introduction of an innovation in 1989-91. Of these, 37 reported a process innovation, 23 an incremental product innovation, and 15 a major product innovation. This indicates that R&D may be more important to the innovation process than the design of the innovation indicator system assumes.

Telecommunication equipment, instruments and machinery: both R&D-intensive and patent-intensive

Annex Table 2 compares R&D intensities at the product group level (comparing R&D with gross production value) and patenting intensity (patent applications per billion markaa of gross production). The most intensive products, in terms of both R&D and patent intensity, are telecommunications equipment and instruments. For other machinery and equipment (electrical and non-electrical), the match also seems quite good. Machine tools have a rather high patenting intensity as compared to R&D intensity. Drugs and petroleum and coal products are quite R&D-intensive, but patenting is comparatively low.

Inter-industry differences in patent trends compared with R&D trends

R&D expenditures have grown almost twice as fast as patent applications during the period studied (see Figure 1), although the situation varies in different industries. Some show an increased propensity to patent. Patent applications related to electrical circuits and telecommunications, for example, have grown very fast, over 700 per cent during the period 1985-92, although R&D expenditure only increased about 100 per cent in constant prices (see Figure 2). Patent applications related to machinery seem to have remained at a constant level, although R&D expenditure has doubled (see Figure 3). A decreasing propensity to patent seems to be the more usual situation. If telecommunications equipment is excluded, total patent applications have remained at the same level, while R&D expenditure has increased by 30 per cent (Figure 4).

◆ Figure 1. ***Trends in R&D and patenting by Finnish firms, 1985-92***

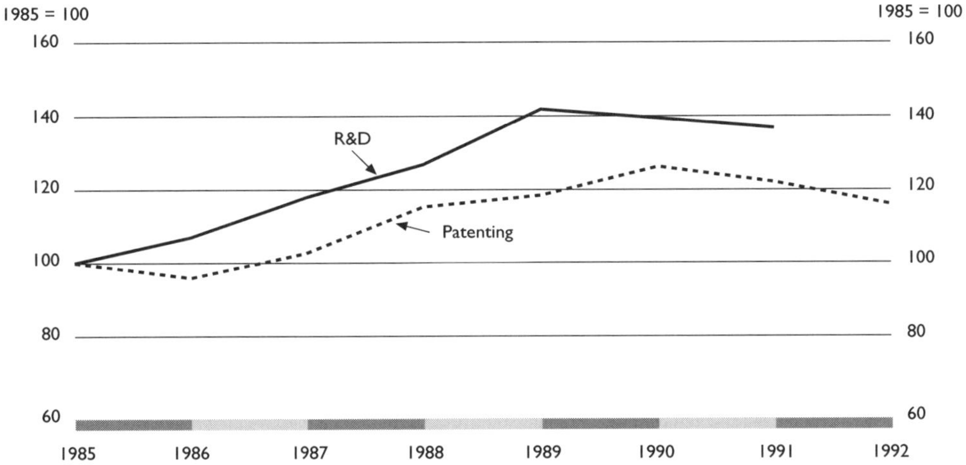

Source: Statistics Finland, *Science and Technology,* 1994.

◆ Figure 2. ***Trends in R&D and patenting, 1985-92***
Telecommunications equipment

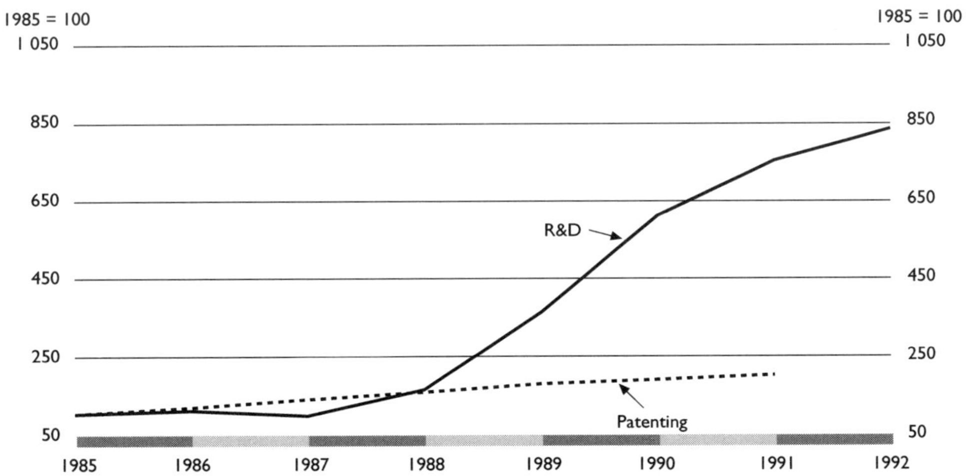

Source: Statistics Finland, *Science and Technology,* 1994.

◆ Figure 3. ***Trends in R&D and patenting, 1985-92 Machinery***

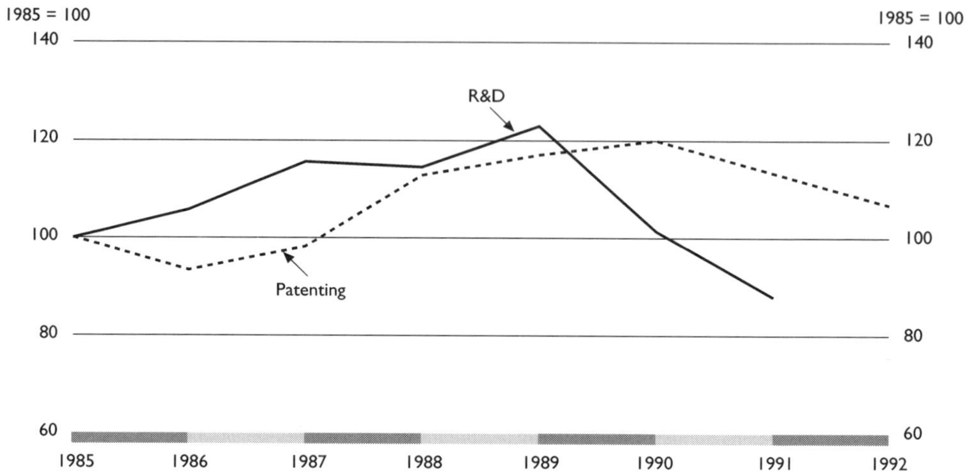

Source: Statistics Finland, *Science and Technology*, 1994.

◆ Figure 4. ***Trends in R&D and patenting by Finnish firms, excluding telecommunications equipment***

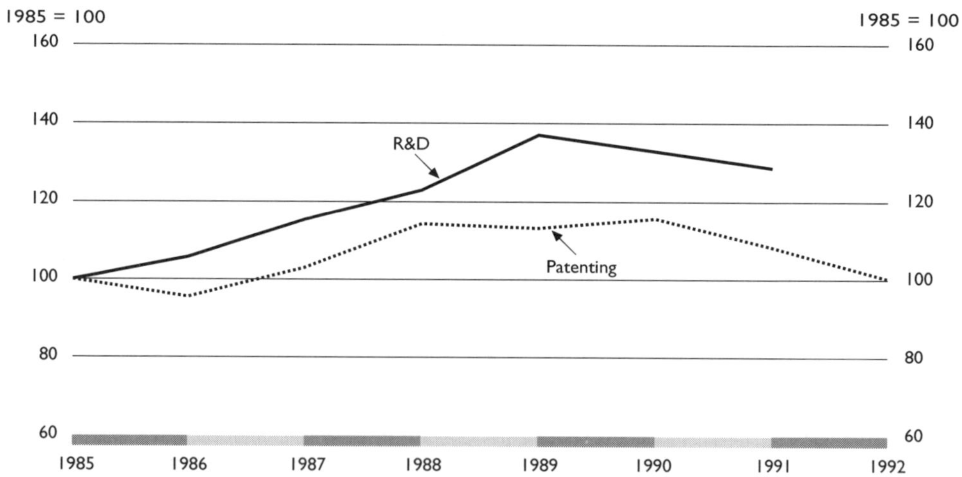

Source: Statistics Finland, *Science and Technology*, 1994.

Only 13 per cent of firms performing R&D applied for a patent in 1992

Annex Table 3 compares the numbers of firms with R&D, the number of firms with patent applications, and the number of firms with both. The data on firms with patent applications and/or R&D were derived from the Finnish patent database and from the data in R&D surveys. The closest connection between patenting and R&D performance was found in the electrotechnical and machinery industries, where the share of firms with R&D that also applied for a patent was higher than average. About 40 per cent of firms that applied for patents in 1992 had not performed R&D in 1991.

Summary and conclusions

It is well known that it is impossible to evaluate the output and impact of R&D with a single measure. This preliminary analysis has shown that the indicator used in innovation surveys, *i.e.* sales due to major and incremental product innovations, and patent applications seem to be good output indicators for industries such as electrotechnical product manufacturing and machinery manufacturing that rely primarily on product innovation and related R&D.

Another group of industries, including textiles, clothes, and furniture, rank rather well in terms of innovative sales, but this cannot be regarded as a result of R&D activity or technological progress, as both R&D intensity and patent intensity are rather low. Input in the form of other innovation expenditure is also rather low. One explanation may be the short life cycle of products, which makes it necessary to renew products quite often without substantial innovation; another may be the design activities included in the innovation process. One solution might be to weight the innovativeness measure with respect to the average life span of the firm's products. The feasibility of collecting this information should be studied.

A third group of industries, including basic metals and pulp and paper, relies heavily on process development. Their innovativeness, in terms of the share of new products, is rather low, but their R&D and patent intensities are relatively high. The problem raised by this apparent mismatch might be solved by requesting information on sales of unchanged products produced by new production methods. The indicator for share of highly innovative firms ranks basic metals and pulp and paper manufacturing industries higher because process innovations are included.

In general, R&D-intensive industries also have high estimated patent intensity and *vice versa*. One explanation for low patent intensity and relatively high R&D intensity might be that activities that modify existing products rather than produce a completely new invention are included in R&D figures. Also, an individual firm's propensity to apply for a patent may vary over time; this must be kept in mind when assessing the usefulness of patent statistics as an output indicator.

In sum, using R&D actitivies as sole indicator of technology inputs does not seem to fully explain the emergence of new products or the development of new inventions. New technology can also be acquired through training and purchase of licences or machinery and result in new products or processes. Correspondingly, the output indicators generally used, new products and patent applications, do not totally cover the results of innovation activity. New indicators of both input and output aspects of innovation activity need to be developed.

Annex

Annex Table 1. **Intensities of R&D and innovativeness, 1991**

(%)

	R&D intensity	Sales due to innovative products
Electrical products	8.1	52.5
Basic metals	2.7	8.6
Machinery	2.3	37.1
Metal products	1.9	12.0
Chemical products	1.5	14.2
Other products	1.5	20.2
Transport equipment	1.4	39.3
Pulp and paper	1.0	13.2
Stone, clay and glass	1.0	32.5
Textiles	0.9	33.4
Food	0.8	12.6
Clothes	0.6	34.0
Furniture	0.4	24.4
Wood and wood products	0.2	16.0
Publishing and printing	0.1	2.3

Source: Statistics Finland.

Annex Table 2. **R&D intensities (R&D expenditure/gross output) and patenting intensities (patent applications/gross output in Mk billion), 1991**

Product group	R&D intensity %	Patent intensity %
Electrical circuits, telecomm. equip.	18.1	30.8
Instruments, optical instruments and photographic equipment	12.3	41.6
Drugs	11.9	8.3
Other machinery and equipment for industry	11.3	23.8
Petroleum and coal products	9.8	1.6
Metallurgical machines	9.8	0.0
Machines for construction and mining	4.6	20.1
Electrical equipment for industry	4.5	12.2
Other chemical products	4.3	2.4
Machines for agriculture and forestry	3.9	15.5
Machine tools	3.8	26.4
Chemicals	3.5	5.0
Pulp and paper making machines	3.3	14.8
Office machines, computers	3.0	10.9
Rubber products	2.2	2.5
Transport equipment	2.1	6.7
Plastic products	1.7	1.5
Fabricated metal products	1.7	10.7
Household machinery and equipment	1.4	9.6
Other manufacturing	1.3	15.8
Non-ferrous metals	1.2	0.3
Pulp, paper and paper products	1.2	1.9
Machines, turbines	1.1	9.5
Other general purpose machinery	1.1	15.0
Ferrous metals	0.9	0.3
Glass, clay, products thereof	0.9	6.8
Textiles, clothes	0.9	0.9
Food, beverages, tobacco	0.8	0.3
Furniture (also of metal and plastics)	0.4	4.2
Wood, products thereof (excl. furniture)	0.3	0.8
Publishing and printing	0.1	0.1
Oil refining	0.0	0.0
Manufacturing total	1.9	4.8

Source: Statistics Finland.

Annex Table 3. **Patenting and R&D: number of firms by industry**

	Number of firms				
	R&D [1]	Patents [2]	R&D, no patents	R&D and patents	Patents, no R&D
Food, beverages	183	12	171	9	3
Textiles,	36	5	31	4	1
Wearing apparel	69	1	68	0	1
Wood and wood products	62	9	53	6	3
Pulp and paper	39	13	36	12	1
Printing and publishing	13		13		
Furniture	88	3	85	2	1
Chemicals	56	11	45	10	1
Petroleum	63	6	59	2	4
Rubber and plastics	12	8	4	5	3
Clay and glass	100	11	89	8	3
Basic metals	12	9	3	6	3
Fabricated metal products	170	24	146	12	12
Machinery	364	70	249	52	18
Electrical products	298	46	252	39	7
Transport equipment	59	17	42	11	6
Other manufacturing	38	11	27	9	2
Other + unknown	553	190	363	99	91
Total firms	2 215	446	1 769	285	161

1. R&D performed in 1991.
2. Patents applied for in 1992.

Source: Statistics Finland.

RELATIONSHIP BETWEEN R&D INPUT AND OUTPUT

by
Christoph Grenzmann
SV-Wissenschaftsstatistik
and
Siegfried Greif
German Patent Office

This chapter reports on a project of the *SV-Wissenschaftsstatistik*, Essen, and the *Deutsches Patentamt* (DPA – German Patent Office), Munich, which studies the relationship between R&D activity and patent activities of companies. The project's objective is to acquire a better understanding of the relationship between R&D input and patent output. To this end, extensive patent and R&D data for companies were collected from inventories of patent applications filed at the German Patent Office and from R&D statistics in the business enterprise sector compiled by SV-Wissenschaftsstatistik. This co-operative project makes it possible to classify the data quantitatively and qualitatively to a degree that would be difficult in any other way.

OBJECTIVES AND PROCEDURES

R&D data

SV-Wissenschaftsstatistik collects data on R&D expenditures in the business enterprise sector in Germany. This information is requested from enterprises and co-operative research institutes every two years. The method of investigation and the evaluation guidelines used are those agreed as international standards within the OECD.

The R&D statistics for the business enterprise sector are a part of the total R&D statistics of Germany, as co-ordinated by the BMFT (Federal Ministry of Research and Technology). As such, they are included in the government's Federal Research Report and in the annual statistical report of the Federal Bureau of Statistics. Evaluations are also undertaken by industrial associations and scientific institutions.

As in other OECD Member countries, these R&D statistics are "input" statistics: the main investigations focus on R&D expenditure and R&D personnel. The following individual areas of investigation are dealt with (Grenzmann, 1993):

- R&D expenditure:
  - by source of funds
  - by objective
- R&D personnel.

It is not possible to make a definitive statement regarding the "efficiency" of R&D activity if input ratings alone are used. It is necessary to set input ratings against the relevant output ratings.

Patent applications

One indicator for evaluating R&D activities is the number of patents or the share of products based on company-owned patents or registered utility models. Patenting data are a very important indicator for economic and technological relationships, especially for the business enterprise sector. The value of the indicator depends to a large extent on the quantity of the data and the ability to relate them to each other.

The patent data are based on patent applications of domestic companies with the German Patent Office. These cover approximately 98 per cent of all registration activities; only about 2 per cent of German patent applications are submitted to the European Patent Office directly, *i.e.* without prior application to the German Patent Office.

When selecting the output rate, a distinction must be made between patents applied for and patents granted. At the German Patent Office, only about one-third of all applications actually become patents. In other words, the number of applications actually includes many more inventions deriving from R&D.

A further advantage of patents applied for is that there is a closer temporal link between R&D and patent applications, as the time that elapses between the R&D and the application is relatively short, while the time between the R&D and the granting of a patent can be very lengthy.

Combining the data

General approach

The relevant data are combined, taking into account the time difference between R&D activities and patent application. The data accumulated from the 1991 R&D surveys and the 1992 inventories (patent applications) of the German Patent

Office were consolidated. Since the data vary in structure and must be treated confidentially, the two institutions have to co-operate closely. A procedure has therefore been worked out to consolidate the two data inventories and, at the same time, meet the needs of confidentiality.

Comparison of data resources

To make it possible to classify the data from the various inventories for the duration of the project, a comparison was carried out to localise and identify every company in the SV address index that had applied for a patent. Comparison of the two data inventories was carried out manually as there is no common system for identifying the addresses.

When comparing the two inventories, particular attention had to be paid to intercompany connections. In practice, companies often transfer their R&D activities to an associated R&D enterprise and make patent applications through another associated enterprise. Thus, although both associated companies are legally independent, patents cannot be applied for by the one in the absence of R&D activities by the other.

The comparison of the two data inventories has revealed (July 1994) 5 100 cases of overlapping, and about 3 400 companies remain to be compared. Future investigations will concentrate initially on the research activities between 1987 and 1991 of about 1 000 companies that filed patent applications at the German Patent Office in 1992. The comparison was based on total inventories and was not organised by branch or size of enterprise. Therefore, so far, the identified enterprises tend to belong to the larger firms. The entities not yet identified are mainly SMEs.

RESULTS

Patent applications

Patent applications by industrial classification

In terms of the major industrial classification of the enterprise, patent applications can be broken down as follows:[1]

- drugs and chemicals: 26.6 per cent;
- machinery: 23.3 per cent;
- electrical industry: 17.2 per cent;
- automobile industry: 14.6 per cent.

These results are, overall, very close to those of previous investigations.[2]

Table 1. **Patent applications 1992 and R&D expenditure 1991 by business enterprises included in the inventories of the German Patent Office and the SV-Wissenschaftsstatistik – by industrial branch and by number of employees**

Enterprises SV with patent applications	German SIC 1	Number of patent applications 1992 2	Share of column 2 (%) 3	R&D expenditure 1991 per patent application 1992 (DM 1 000) 4
Branches of industry				
Electricity, gas and water supply, mining	1	121	0.87	3 824.07
Electricity, gas and water supply	10	42	0.30	3 202.36
Mining	11	79	0.57	4 154.61
Manufacturing	2	13 600	98.20	2 773.21
Chemicals and chemical products	20	3 700	26.72	2 549.32
Chemicals and drugs	200	3 689	26.64	2 527.10
Rubber and plastic products	21	331	2.39	1 271.98
Plastic products	210	200	1.44	1 098.19
Rubber	213	131	0.95	1 537.31
Stone, clay, glass (non-metallic products)	22	214	1.55	1 212.61
Basic metals	23	484	3.49	889.96
Ferrous metals	230-232	272	1.96	1 003.08
Non-ferrous metals	233	115	0.83	911.98
Fabricated metal products, construction machinery,	24	5 848	42.23	3 256.28
computing machinery	242	3 220	23.25	975.74
Motor vehicles	244	2 015	14.55	4 901.04
Aerospace	248	330	2.38	11 841.79
	25	2 912	21.03	2 732.66
Electrical machinery	250	2 376	17.16	2 933.42
Medical, precision and optical instruments	252-254	297	2.14	1 861.50
Watches and clocks (instruments)	256	222	1.60	1 705.71
Wood, paper, printing, publishing	26	52	0.38	1 087.83
Wood (and furniture)	260/261	16	0.12	1 318.69

Table 1. **Patent applications 1992 and R&D expenditure 1991 by business enterprises included in the inventories of the German Patent Office and the SV-Wissenschaftsstatistik – by industrial branch and by number of employees** *(cont'd)*

Enterprises SV with patent applications	German SIC 1	Number of patent applications 1992 2	Share of column 2 (%) 3	R&D expenditure 1991 per patent application 1992 (DM 1 000) 4
Textiles, wearing apparel, fur and leather	27	39	0.28	833.10
Textiles	275	23	0.17	936.83
Food, beverages and tobacco	28/29	20	0.14	4 130.75
Agriculture, service sector, construction	0, 3-7	128	0.92	3 472.73
Total		13 849	100.00	2 788.85
Size (No. of employees)				
0-99		305	2.20	1 000.32
100-499		1 051	7.59	776.79
500-999		645	4.66	1 312.72
1 000-1 999		1 177	8.50	1 857.27
2 000-4 999		2 198	15.87	1 718.27
5 000-9 999		2 255	16.28	2 202.73
= > 10 000		6 218	44.90	4 137.13
Total		13 849	100.00	2 788.85

Basis: R&D data and patent applications for individual enterprises.
Source: SV-Wissenschaftsstatistik/German Patent Office.

Patent applications by size of company

The breakdown of patent applications by company size is affected by the fact that the comparison of inventories is incomplete, with the result that smaller and medium-sized companies are insufficiently represented. Nonetheless, several tendencies are apparent.

A categorisation of patent applications by company size (see Table 1) shows that major concerns (those with more than 10 000 employees) account for about 45 per cent of total patent applications. Previous surveys found a similar concentration (Greif, 1993). Patent applications are divided among the various company categories. The same is true for individual classifications by sector of industrial activity (Table 2).[3]

Research and development

In 1991, the approximately 1 000 business enterprises included in the above survey spent DM 38.6 billion on R&D. Given that the industrial sector spent a total of about DM 56.3 billion on R&D (Grenzmann *et al.*, 1994), the enterprises surveyed accounted for around 69 per cent of R&D expenditure. Thus, it is possible to draw some general conclusions, even though the comparison of enterprises is incomplete.

R&D and patent applications

In terms of industrial classification, patent applications and R&D expenditure of companies (see Table 1) show patterns characteristic of their branch. Individual branches have the following rates of expenditure on R&D per patent application:

- mechanical engineering: ca. DM 1.0 million;
- chemical industry: ca. DM 2.5 million;
- automobile industry: ca. DM 4.9 million;
- aerospace industry: ca. DM 12.8 million.

Obviously, the relation between expenditure on R&D and patent applications varies greatly. The figures indicate both the level of R&D expenditure required for an invention and the extent to which patentable inventions are registered for patenting. In machinery, for instance, the figures are relatively low, whereas they are relatively high in the automotive and aerospace industries.[4] However, the relatively small number of patents in the automotive industry is due to the fact that a large share of R&D expenditure is spent on styling (optimisation in wind tunnel) and testing of new models, activities which produce relatively few patentable inventions. The similar situation in the aerospace industry is the result of the importance given to testing prototypes and new models.[5]

Table 2. **Patent applications 1992 and R&D expenditure 1991 by business enterprises included in the inventories of the German Patent Office and SV-Wissenschaftsstatistik**

By selected industrial branches according to number of employees

Enterprises with patent applications Size (No. of employees)	Number of patent applications 1992 1	Share of column 1 (%) 2	R&D expenditure 1991 per patent application 1992 (DM 1 000) 3
Chemicals and drugs			
0-99	11	0.30	1 002.09
100-499	38	1.03	1 678.24
500-999	39	1.06	3 350.41
1 000-1 999	278	7.54	2 235.52
2 000-4 999	239	6.48	2 366.90
5 000-9 999	484	13.12	3 113.56
= > 10 000	2 600	70.48	2 470.33
Total	3 689	100.00	2 527.10
Machinery, computing machinery			
0-99	82	2.55	287.01
100-499	440	13.66	695.38
500-999	292	9.07	1 097.97
1 000-1 999	389	12.08	1 074.45
2 000-4 999	990	30.75	928.70
5 000-9 999	729	22.64	904.23
= > 10 000	298	9.25	1 661.76
Total	3 220	100.00	975.74
Motor vehicles			
0-99	3	0.15	181.43
100-499	22	1.09	586.73
500-999	54	2.68	369.52
1 000-1 999	93	4.62	892.85
2 000-4 999	222	11.02	2 699.80
5 000-9 999	428	21.24	2 099.10
= > 10 000	1 193	59.21	6 924.88
Total	2 015	100.00	4 901.04
Electrical machinery			
0-99	85	3.58	507.25
100-499	181	7.62	1 257.82
500-999	82	3.45	2 186.39
1 000-1 999	147	6.19	2 157.66
2 000-4 999	336	14.14	1 929.24
5 000-9 999	287	12.08	2 666.10
= > 10 000	1 258	52.95	3 806.96
Total	2 376	100.00	2 933.42
Medical, precision and optical instruments, etc.			
0-99	12	4.04	338.33
100-499	55	18.52	591.36
500-999	20	6.73	1 911.30
1 000-1 999	21	7.07	1 871.67
2 000-4 999	84	28.28	2 182.01
5 000-9 999	88	29.63	2 739.34
= > 10 000	17	5.72	846.88
Total	297	100.00	1 861.50

Basis: R&D data and patent applications for individual enterprises.
Source: SV-Wissenschaftsstatistik/German Patent Office.

Table 3. **International patent classifications included in the survey**

Section A – Human necessities	
Subsection:	Agriculture
A 01	Agriculture; forestry; animal husbandry; hunting; trapping; fishing
Subsection:	Personal or domestic articles
A 47	Furniture; domestic articles or appliances; coffee mills; spice mills; suction cleaners in general
Subsection:	Health; amusement
A 61	Medical or veterinary science; hygiene
A 62	Life-saving; fire-fighting
Section B – Performing operations; transporting	
Subsection:	Separating; mixing
B 01	Physical or chemical processes or apparatus in general
B 05	Spraying or atomising in general; applying liquids or other fluid materials to surfaces, in general
B 08	Cleaning
Subsection:	Shaping
B 21	Mechanical metalworking without essentially removing material; punching metal
B 22	Casting; powder metallurgy
B 23	Machine tools; metalworking not otherwise provided for
B 25	Hand tools; portable power-driven tools; handles for hand implements; workshop devices; manipulators
B 29	Working of plastics; working of substances in a plastic state in general
Subsection:	Printing
B 41	Printing; lining machines; typewriters, stamps
B 42	Bookbinding; albums; files; special printed matter
Subsection:	Transporting
B 60	Vehicles in general
B 62	Land vehicles for travelling otherwise than on rails
B 64	Aircraft; aviation; cosmonautics
B 65	Conveying; packing; storing; handling thin or filamentary material
B 66	Hoisting; lifting; hauling
Section C – Chemistry; metallurgy	
Subsection:	Chemistry
C 01	Inorganic chemistry
C 03	Glass, mineral, or slag wool
C 04	Cement; concrete; artificial stone; ceramics; refractories
C 07	Organic chemistry
C 08	Organic macromolecular compounds; their preparation or chemical working-up; compositions based thereon
C 09	Dyes; paints; polishes; natural resins; adhesives; miscellaneous compositions; miscellaneous applications of materials
C 10	Petroleum, gas or coke industries; technical gases containing carbon monoxide; fuels; lubricants; peat
C 11	Animal or vegetable oils, fats, fatty substances or waxes; fatty acids therefrom; detergents; candles
C 12	Biochemistry; beer; spirits; wine; vinegar; microbiology; enzymology; mutation or genetic engineering
Subsection:	Metallurgy
C 23	Coating metallic material; coating material with metallic material; chemical surface treatment; diffusion treatment of metallic material; coating by vacuum evaporation, by sputtering, by ion implantation or by chemical vapour deposition, in general; inhibiting corrosion of metallic material or incrustation in general
C 25	Electrolytic or electrophoretic processes; apparatus therefor

Table 3. **International patent classifications included in the survey** *(cont'd)*

Section D – Textiles; paper	
Subsection:	Textiles or flexible materials not otherwise provided for
D 01	Natural or artificial threads or fibres; spinning
D 02	Yarns; mechanical finishing of yarns or ropes; warping or beaming
D 06	Treatment of textiles or the like; laundering; flexible materials not otherwise provided for
Subsection:	Paper
D 21	Paper-making; production of cellulose
Section E – Fixed constructions	
Subsection:	Building
E 05	Locks; keys; window or door fittings; safes
Section F – Mechanical engineering; lighting; heating; weapons; blasting	
Subsection:	Engines or pumps
F 01	Machines or engines in general; engine plants in general; steam engines
F 02	Combustion engines; hot-gas or combustion-product engines plants
F 04	Positive-displacement machines for liquids; pumps for liquids or elastic fluids
Subsection:	Engineering in general
F 15	Fluid-pressure actuators; hydraulics or pneumatics in general
F 16	Engineering elements or units; general measures for producing and maintaining effective functioning of machines or installations; thermal insulation in general
F 17	Storing or distributing gases or liquids
Subsection:	Lighting; heating
F 21	Lighting
F 28	Heat exchange in general
Subsection:	Weapons; blasting
F 41	Weapons
F 42	Ammunition; blasting
Section G – Physics	
Subsection:	Instruments
G 01	Measuring; testing
G 02	Optics
G 03	Photography; cinematography; analogous techniques using waves other than optical waves; electrography; holography
G 05	Controlling; regulating
G 06	Computing; calculating; counting
G 08	Signalling
G 11	Information storage
Section H – Electricity	
H 01	Basic electrical elements
H 02	Generation, conversion, or distribution of electric power
H 03	Basic electronic circuitry
H 04	Electric communication techniques
H 05	Electric techniques not otherwise provided for

Table 4. **Patent applications 1992 and R&D expenditure 1991**
German Patent Office and
By industrial branch, number of employees

Enterprises SV with patent applications	German SIC				A61	B01	B23
		1			2	3	4
Branches of industry							
Electricity, gas and water supply, mining	1				14		1
Electricity, gas and water supply		10			6		
Mining		11			8		1
Manufacturing	2			367	272	213	259
Chemicals and chemical drugs		20		218	96	8	61
Chemical drugs			200	218	95	8	61
Rubber and plastic products		21		12	3	4	33
Plastic products			210	6	2	3	13
Rubber			213	6	1	1	20
Stone, clay, glass (non-metallic mineral products)		22		4	12	1	2
Basic metals		23		1	29	16	6
Ferrous metals			230-232		8	14	4
Non-ferrous metals			233		12	1	
Fabricated metal products, construction machinery,		24		15	111	149	145
computing machinery			242	6	54	112	120
Motor vehicles			244	4	47	27	18
Aerospace			248	1		7	3
		25		109	21	34	12
Electrical machinery			250	32	10	26	10
Medical, precision and optical instruments,			252-254	71	11	1	
Watches and clocks (instruments)			256	6		7	2
Wood, paper, printing, publishing		26					
Wood (and furniture)			260/261				
Textiles, wearing apparel, fur and leather		27		7			
Textiles			275	7			
Food, beverages and tobacco		28/29		1		1	
Agriculture, service sector, construction	0, 3-7				9	1	1
Total				367	295	214	261
Size (No. of employees)							
< 100				21	17	5	2
100-499				38	45	36	45
500-999				20	12	18	17
1 000-1 999				65	22	22	30
2 000-4 999				88	49	30	47
5 000-9 999				50	28	24	21
= > 10 000				85	122	79	99
Total				367	295	214	261

Basis: R&D-data and patent applications for individual enterprises.
Source: SV-Wissenschaftsstatistik/German Patent Office

by business enterprises included in the inventories of the the SV-Wissenschaftsstatistik

and selected international patent classifications

Number of patent applications 1992 in selected IPC classes											
B29	B41	B60	B65	C07	F02	G01	H01	H04	Sub-total 2-13	Column 16 – column 14	Total
5	6	7	8	9	10	11	12	13	14	15	16
1		7	1	2	11		2		39	82	121
		1	1	2	3				13	29	42
1		6			8		2		26	53	79
305	1 133	556	1 223	389	659	526	343		6 245	7 355	13 600
13	10	60	1 212	2	60	12	3		1 755	1 945	3 700
13	10	60	1 211	2	60	12	3		1 753	1 936	3 689
5	41	24		10	4	1			137	194	331
	4	17		10		1			56	144	200
5	37	7			4				81	50	131
2	2	5		1	6	15	1		51	163	214
32	16	15	2		22	13	2		154	330	484
31	5	10			18	6	2		98	174	272
1	2	1	2		4	4			27	88	115
248	854	383	5	230	245	122	36		2 543	3 305	5 848
238	155	347	5	72	122	29	6		1 266	1 954	3 220
1	681	16		150	67	48	9		1 068	947	2 015
	11	1		5	52	39	7		126	204	330
3	208	51	3	146	314	363	301		1 565	1 347	2 912
	177	27		146	247	353	298		1 326	1 050	2 376
1	1	6	2		61	8	3		165	132	297
2	30	16	1		6	1			71	151	222
1		15			4				20	32	52
		3			1				4	12	16
1	2	3			4				17	22	39
1	1	1							10	13	23
			1						3	17	20
1	4	5			18				39	89	128
307	1 137	568	1 224	391	688	526	345		6 323	7 526	13 849
1	2	26	2	1	38	13	4		132	173	305
8	23	64	5	3	88	40	25		420	631	1 051
5	35	40	4	2	34	21	1		209	436	645
63	83	75	65	9	48	62	25		569	608	1 177
7	183	154	90	50	89	97	75		959	1 239	2 198
177	199	120	141	51	116	67	6		1 000	1 255	2 255
46	612	89	917	275	275	226	209		3 034	3 184	6 218
307	1 137	568	1 224	391	688	526	345		6 323	7 526	13 849

Table 5. **R&D expenditure 1991 per patent application 1992 and R&D of the German Patent Office**

By industrial branch, number of employees and

Enterprises SV with patent applications	German SIC 1			R&D expenditure 1991 (in DM 1 000) A61 2	B01 3	B23 4	B29 5	B41 6
Branches of industry	1			4 887.31		504.50	23.33	
Electricity, gas and water supply, mining		10			2 210.38			
Electricity, gas and water supply		11			6 895.02		504.50	23.33
Mining	2			2 915.48	1 958.20	3 064.40	1 684.85	955.12
Manufacturing		20		3 438.33	2 896.72	3 051.77	2 439.35	2 286.19
Chemicals and chemical drugs			200	3 438.33	2 176.68	3 051.77	2 439.35	2 286.19
Chemical drugs		21		1 076.47	2 054.65	853.37	1 355.77	1 169.90
Rubber and plastic products			210	621.21	2 519.38	737.92	1 717.12	
Plastic products			213	1 531.74	1 125.19	1 199.71	1 120.89	1 169.90
Rubber		22		483.01	971.58	1 626.00	716.62	1 242.17
Stone, clay, glass (non-metallic mineral products)		23		80.00	750.59	2 466.63	165.20	174.94
Basic metals			230-232		171.01	2 589.25	171.01	171.01
Ferrous metals			233		1 489.13	2 591.67		296.75
Non-ferrous metals		24		11 317.36	1 530.72	3 173.40	1 535.21	915.51
Fabricated metal products, construction machinery, computing machinery			242	1 741.69	1 020.79	1 754.98	889.22	615.19
Motor vehicles			244	3 119.13	2 039.62	6 617.05	3 200.32	1 270.00
Aerospace			248	12 071.83		12 947.71	12 071.83	
		25		1 169.39	2 145.04	3 253.19	1 483.77	6 541.60
Electrical machinery			250	908.62	1 674.13	3 938.58	1 474.09	
Medical, precision and optical instruments,			252-254	1 368.60	2 573.14	3 562.22		2 144.79
Watches and clocks (instruments)			256	202.88		663.30	1 532.17	8 740.00
Wood, paper, printing, publishing		26						662.50
Wood (and furniture)			260/261					
Textiles, wearing apparel, fur and leather		27		641.14				326.08
Textiles			275	641.14				326.08
Food, beverages and tobacco		28/29		3 783.75		351.00		
Agriculture, service sector, construction	0, 3-7				992.29	1 059.68	1 162.50	1 335.00
Total				2 915.48	2 067.74	3 055.03	1 678.32	953.32
Size (No. of employees)								
< 100				293.69	2 646.50	930.80	245.00	1 335.00
100-499				761.45	619.94	1 304.38	483.58	377.56
500-999				3 895.09	2 869.86	1 540.38	497.11	708.50
1 000-1 999				3 174.26	1 129.89	2 654.01	1 205.04	247.63
2 000-4 999				2 004.32	2 903.67	1 486.51	1 341.20	4 516.13
5 000-9 999				5 829.74	1 758.87	3 591.06	1 604.41	722.46
= > 10 000				3 326.86	2 346.49	4 876.81	2 772.32	2 384.43
Total				2 915.48	2 067.74	3 055.03	1 678.32	953.32

Basis: R&D-data and patent applications for individual enterprises.
Source: SV-Wissenschaftsstatistik/German Patent Office.

expenditure 1991 by business enterprises included in the inventories and the SV-Wissenschaftsstatistik

selected international patent classifications

per patent application 1992 in selected IPC classes

B60 7	B65 8	C07 9	F02 10	G01 11	H01 12	H04 13	Column 2-13 14	Other classes 15	Total 16
5 712.00	4 959.45	847.73	4 092.95		180.00	4 127.47		3 824.07	
	4 959.45	4 959.45	847.73	3 386.30			2 695.04		3 202.36
	5 837.42			4 357.94		180.00	4 843.69		4 154.61
2 921.45	1 414.65	2 733.05	3 449.78	4 011.86	4 395.92	4 027.48	2 898.57		2 773.21
1 615.54	1 995.96	2 732.32	2 621.83	3 784.29	3 598.09	1 927.10	2 825.82		2 549.32
1 615.54	1 995.96	2 732.01	2 621.83	3 784.29	3 598.09	1 927.10	2 786.60		2 527.10
2 208.84	1 153.13		1 398.75	1 414.56	1 398.75		1 550.12		1 271.98
1 080.81	1 020.35		1 398.75		1 398.75		1 256.39		1 098.19
2 330.79	1 475.59			1 414.56			1 753.20		1 537.31
2 060.10	1 041.08		1 626.00	1 145.25	1 317.06	552.46	1 122.86		1 212.61
597.09	2 344.60	296.75		303.01	594.26	171.01	830.86		889.96
167.70	3 365.51			318.91	636.46	171.01	897.93		1 003.08
490.50	296.75	296.75		231.48	566.00		956.26		911.98
3 267.71	1 241.00	1 121.10	4 522.26	6 021.29	7 014.95	12 318.97	3 282.87		3 256.28
874.75	857.23	1 121.10	1 885.00	1 038.25	1 087.88	6 443.55	1 016.19		975.74
3 764.84	2 594.96		5 602.06	10 153.13	5 039.44	6 106.15	4 465.05		4 901.04
7 792.28	12 071.83		12 552.14	12 063.68	12 016.78	12 071.83	11 745.54		11 841.79
1 904.84	1 992.39	3 360.72	1 924.58	2 911.73	3 813.71	3 093.90	2 772.75		2 732.66
1 990.53	2 149.17		1 924.58	3 154.68	3 817.21	3 103.06	2 945.47		2 933.42
72.33	2 093.67	4 275.00		1 910.72	3 346.70	2 184.63	1 831.81		1 861.50
1 460.36	1 744.68	1 532.17		3 086.94	1 561.40		1 686.56		1 705.71
	1 288.39			1 291.83			1 257.79		1 087.83
	3 389.39			188.00			2 589.04		1 318.69
1 388.21	833.27			724.91			764.11		833.10
2 051.50	1 050.00						791.56		936.83
		14 670.00					6 268.25		4 130.75
18 562.86	770.44			7 778.85			5 913.10		3 472.73
2 976.48	1 461.94	2 734.87	3 436.48	4 111.71	4 395.92	4 005.17	2 924.74		2 788.85
202.15	1 000.70	3 723.00	1 626.00	1 562.83	601.01	299.50		829.86	1 000.32
295.10	718.59	1 866.57	3 209.78	912.36	915.66	807.24		759.11	776.79
944.62	828.01	2 836.92	2 697.68	2 042.18	1 172.12	6 000.00		1 173.56	1 312.72
2 208.52	1 128.44	3 033.74	3 413.24	4 297.61	1 620.10	2 269.82		1 620.64	1 857.27
911.05	1 091.21	2 801.42	1 164.22	2 672.23	2 591.54	2 193.76		1 630.36	1 718.27
1 880.61	1 208.56	3 114.21	3 463.39	4 298.17	2 105.57	3 309.95		2 059.08	2 202.73
4 280.61	3 680.33	2 650.96	3 859.81	6 098.36	7 744.70	5 326.64		4 210.41	4 137.13
2 976.48	1 461.94	2 734.87	3 436.48	4 111.71	4 395.92	4 005.17		2 674.69	2 788.85

The breakdown of patent data by size of enterprise clearly shows that smaller and medium-sized companies conduct more intensive R&D with a view to patenting. The larger the company, the lower the number of patent applications in proportion to the amount of R&D expenditure. In absolute figures, this results in an average R&D expenditure per patent application of DM 1 million for the smallest company category and of DM 4.1 million for the largest company category. In addition to this general pattern, individual branches also show specific variances (see Table 2).

Patent applications according to IPC by industrial classification and company size

For this presentation IPC class levels have been used.[6] A breakdown of patent applications by International Patent Classification (IPC) and by the industrial classification (SIC) of the patenting company shows that, on the basis of the IPC categories with the highest ratings (see Table 4), the dispersion of inventions among individual IPC classification units varies according to branch. In the chemicals industry, for instance, 32 per cent of patents concern IPC category C07 (organic chemistry), whereas in electrical engineering this share is only achieved for a group of three IPC categories.

With regard to company size, various patterns became obvious. Large-scale enterprises carry out patenting in IPC category C07 (organic chemistry), for example, while more medium-sized companies apply for patents in IPC category B29 (plastic processing).

In terms of R&D expenditure per patent application, it is apparent that the high rates in the aerospace and chemicals industries are unrelated to the specific IPC class, while in machinery, for example, the rates vary marked in the different classes (see Table 5). In machinery, the rates vary between DM 0.89 million and DM 1.755 million, while in the case of aerospace, the expenditure per patent in all IPC classes is around DM 12 million, the only exception being B60 (vehicles in general).

FUTURE WORK

Besides expanding the comparison of enterprises in the inventories of the R&D survey and the German Patent Office to increase the share of patent applications that can be attributed to research companies, the following steps are scheduled to be taken.

The R&D survey is based on an analysis of companies by industrial classification. During various discussions in connection with the R&D statistics, the question was raised as to whether R&D activities were to be classified by sector of industrial

Table 6. **Table of concordance: German Standard Industrial Classification (WZ 79) and those of ISIC and NACE**

German Standard Industrial Classification WIRTSCHAFTSZWEIGE			English translation Branch of industry	ISIC Rev. 2 approximate	NACE Rev. 1
1		Energie- u. Wasserversorgung, Bergbau	Electricity, gas and water supply, mining	2 + 4	
	10	Elektriz.-, Gas-, Fernwärme- u. Wasservers.	Electricity, gas and water supply	4	40 + 41
	11	Bergbau	Mining	2	10-14
2		Verarbeitendes Gewerbe	(Manufacturing)	3	15-37
	20	Chem. Industrie usw., Mineralölverarb.	Chemicals and chemical products	351 + 352	24
	200	Chemische Industrie	Chemicals and drugs	351 + 3522	
	21	H. v. Kunststoff- u. Gummiwaren	Rubber and plastic products	355 + 356	25
	210	H. v. Kunststoffwaren	Plastic products	–	–
	213	H. v. Gummiwaren	Rubber	–	–
	22	Gew. u. Verarb. v. Steinen u. Erden usw.	Stone, clay, glass (non-metallic mineral products)	36	26
	23	Metallerzeugung u. -bearbeitung	Basic metals	37	27
	230-232	Eisenschaffende Industrie	Ferrous metals	371	
	233	NE-Metallerzeugung usw.	Non-ferrous metals	372	
	24	Stahl-, Maschinen- u. Fahrzeugbau usw.	Fabricated metal products, construction	381	28
	242	Maschinenbau	Machinery, computing machinery	382	29
	244	H. v. Kraftwagen u. deren Teilen	Motor vehicles	3843	34
	248	Luft- u. Raumfahrzeugbau	Aerospace	3845 + 3829 (part)	35.3
	25	E-technik, Feinmech., H. v. EBM-Waren	Electrical machinery, etc.		
	250	Elektrotechnik	Electrical machinery	383	31
	252-254	Feinmechanik, Optik	Medical, precision and optical instruments, Watches and clocks (instruments)	385	33
	256	H. v. EBM-Waren	Tools and finished metal goods	–	–
	26	Holz-, Papier- u. Druckgewerbe	Wood, paper, printing, publishing	331 + 34 + 3832	20-22
	260/261	Holzbearbeitung, Holzverarbeitung	Wood (and furniture)	331 + 332	20-22, 36.1
	27	Leder,- Textil- u. Bekleidungsgewerbe	Textiles, wearing apparel, fur and leather	32	17-19
	275	Textilgewerbe	Textiles	321	17
	28/29	Ernährungsgewerbe, Tabakverarbeitung	Food, beverages and tobacco	31	15 + 16
0, 3-7		Restliche Wirtschaftsabteilungen	Agriculture, service sector, construction	1, 4, 5, 6-9	01, 02, 05, 45, 0-99

Source: Statistisches Bundesamt, Wiesbaden.

activity or product group for the R&D expenditure or R&D personnel indicators. The R&D survey of the German business enterprise sector aims not only to obtain an analysis by industrial classification, but also an analysis of the applied R&D expenditure by product groups, using a rough index of 33 selected product groups which fits the German product group classification.[7]

Since the patent data (German Patent Office) can be connected to data on companies and R&D (SV-Wissenschaftsstatistik), it is also possible to establish a concordance between the IPC and the product groups (see also Grupp, 1994). This facilitates making connections both between the IPC and the industrial classification and between the IPC and the product groups. While in the former case deductions can be made concerning patent applications and industrial orientation, the latter provides information on the relationship between patent application and research goals.

So far, the study has been based on patent applications in 1992 and R&D expenditure in 1991. This choice is motivated by the fact that R&D expenditure leads to inventions, or to patent applications, in the short term.[8] In the expanded study, the research expenditure of earlier years will be included in the survey to gain more knowledge about the time between research activities and patent application in individual enterprises. The survey will also be extended to include R&D personnel. This will indicate the extent to which patent applications in certain companies are connected with intensive research by individual staff members.

NOTES

1. The tables presented in this paper follow the German Standard Industrial Classification (WZ 79). For international comparison there is a table of concordance, providing the relation of the German Standard Industrial Classification (SIC) to the International Standard Industrial Classification (ISIC) and NACE (Table 6).
2. Grief (1993), p. 52 ff.
3. Previous studies have accorded a higher share to smaller and medium-sized enterprises (see Greif, 1993).
4. These constellations correspond, to a large extent, to the results of an analysis of branch relationships in the United States by Scherer (1983).
5. Interesting relationships can also be determined by using total figures instead of figures on individual companies. In this case, the 1991 R&D expenditure figures amounting to around DM 56 billion would mean about 28 000 patent applications by domestic companies in 1992, or an average rate of about DM 2 million a year for each patent application.
6. At a later date, the enquiry will be expanded to IPC sub-classes.
7. Statistiches Bundesamt (1989), *Systematisches Güterverzeichnis für Produktionsstatistiken*, Stuttgart.
8. A survey for Germany concludes that the period of time between R&D activities and patent applications is between one and two years (Greif, 1985).

REFERENCES

GREIF, S. (1985), "Relationship Between R&D Expenditure and Patent Applications", *World Patent Information*, No. 3.

GREIF, S. (1993), "Patente als Indikatoren für Forschungs-und Entwicklungstätigkeiten", in SV-Wissenschaftsstatistik (ed.), *Forschung und Entwicklung in der Wirtschaft*, Materialien zur Wissenschaftsstatistik, Essen.

GRENZMANN, Chr. (1993), "Methodik und Aufbau der deutschen FuE-Statistik und Struktur der FuE-Aktivitäten in der Bundesrepublik Deutschland", in SV-Wissenschaftsstatistik (ed.), *Forschung und Entwicklung in der Wirtschaft*, Materialien zur Wissenschaftsstatistik, Essen.

GRENZMANN, Chr., R. MARQUARDT, Chr. REVERMANN and J. WUDTKE (1994), *Forschung und Entwicklung in der Wirtschaft 1991 – mit ersten Daten bis 1993*, SV-Wissenschaftsstatistik, Essen.

GRUPP, H. (1994), "Innovationsaktivitäten in strategischen Sektoren: Meßmethodik und empirische Befunde für ausgewählte Technologiegebiete", in SV-Wissenschaftsstatistik, *Wissenschafts- und Technologieindikatoren, Technologische Indikatoren*, Essen.

SCHERER, F. M. (1983), "The Propensity to Patent", *International Journal of Industrial Organization*, No. 1, p. 107 ff.

4

INNOVATION IN FRENCH MANUFACTURING INDUSTRY: A REVIEW OF THE FINDINGS OF THE COMMUNITY INNOVATION SURVEY[1]

by

S. Lhuillery

Centre de recherche en économie industrielle, Université Paris-Nord

INTRODUCTION

The second French innovation survey was conducted in 1993 on the basis of the European questionnaire. The questionnaire, which was not mandatory, was sent to a representative sample of over 4 500 industrial firms with more than 20 employees which had replied to the 1991 Innovation Survey conducted by SESSI *(Service des Statistiques industrielles).* The response rate was approximately 82 per cent, with 3 880 replies that could be processed. Three factors explain why the response rate was so high: the brevity of the questionnaire (two pages), the reference to the forecasts made by the same firms in 1991, and, lastly, the fact that after each survey each firm is notified of its position in its particular sector of activity. The sample was adjusted on the basis of data from the SESSI Annual Survey of Firms, bringing its coverage to a total of some 23 709 firms representative of manufacturing industry as a whole.

The purpose of this article is to review the initial findings from these data. The main findings will be presented in accordance with the French questionnaire (see Box). The general approach adopted here contrasts with papers of a more academic nature. Points of theory are not addressed directly, but two aspects are explored in detail: a new performance variable, namely the ability of firms to remain innovative, and the means used by firms in order to acquire new technology, particularly the R&D links established with their different partners.

INNOVATION IN FRENCH INDUSTRY

Propensity to innovate

Slightly less than 40 per cent of firms stated that they had developed technological innovations during the period 1990-92. Large firms innovated the most. 90 per cent of firms with over 2 000 employees developed innovations over the

COMMUNITY INNOVATION SURVEY – CIS – FOR FRANCE: METHODOLOGY

Sampling and adjustment methods

The CIS for France is one of three surveys conducted by SESSI in 1993, within the framework of co-operation between SESSI and the French Ministry of Higher Education and Research. The themes of the other two surveys were technology appropriation and the organisational structure of firms.

The CIS questionnaire is based on the OSLO Manual which sets out to standardise the collection of technological innovation data. A number of European countries have conducted similar surveys on this basis since 1991.

The French questionnaire was sent to a representative sample of more than 4 500 industrial firms with more than 20 employees which had replied to the 1991 SESSI Innovation Survey. All French firms employing more than 1 000 persons were consulted, half of those with between 500 and 1 000 employees and a third of those with between 20 and 500 employees.

The response rate was approximately 82 per cent, yielding 3 880 replies that could be processed. The results were adjusted on the basis of 1992 data from the SESSI Annual Survey of Firms using strata corresponding to the sectors denoted in the NACE (General Industrial Classification of Economic Activities within the European Communities) by two-digit codes with seven size categories. The adjusted sample accordingly represents a population of some 23 709 firms representative of the whole of manufacturing industry. The adjustment brings the overall innovation rate down from 43 per cent to 39 per cent, confirming the importance of such an operation. The reliability of the results is strengthened by the fact that they differ only slightly (by less than 1 per cent) from those obtained by the parallel "Yale 2" survey. Although some bias may have been introduced by non-response to the questionnaire (18 per cent or so of firms), this bias was tested and found to be insignificant at the time of the previous survey.

Definitions

A firm is classed as innovative when it has developed or introduced technologically innovative products or processes during the period under review.

A product is considered to be technologically innovative if it leads to the creation of a new market or if it differs substantially from products manufactured previously, in terms of either embodied technology or the services offered to users. This definition excludes purely aesthetic or stylistic innovations (design), but does include innovations in packaging or packing.

A process is considered to be technologically innovative if it employs new techniques for the production of products that are either innovative or part of the existing product range of the firm.

(continued on next page)

(continued)

The questionnaire

The firms were asked to supply information regarding:

- the aims of their innovation efforts;
- internal and external sources of innovation;
- the means used to acquire technologies;
- the means used for the diffusion of technology;
- their partners in R&D;
- the reasons for these R&D links;
- the share of their turnover and exports generated by innovative products introduced within the last three years;
- their innovation plans for the next three years.

period under review, but in half the firms with between 20 and 90 employees, a category which accounts for approximately 80 per cent of all industrial firms, did not innovate.

There are also wide sectoral differences. The most innovative sectors are computers (64 per cent), instruments (63 per cent) and electrical machinery (60 per cent). On the other hand, the level of innovation in traditional manufacturing sectors, such as wood, leather, footwear and clothing, is very low.

96 per cent of large firms employing more than 2 000 persons innovated during the period 1986-90. Although this figure dropped to 90 per cent for the following three years, suggesting that the propensity to innovate was adversely affected by the recession, direct comparison with the industrial growth years is dangerous since the length of the reference period is not the same (see Box). Nonetheless, the fact remains that innovation in products and processes concerns 29 per cent and 27 per cent of firms respectively, compared with 53 per cent and 45 per cent during the period 1986 to 1990. The particularly sharp drop in product innovation seems accurately to reflect the sluggishness of markets, while process innovations, which continue to raise productivity levels, are less affected. The relative decline in product innovation is consistent with the predictions made by the firms back in 1990, when they were already planning to focus their efforts on improving the efficiency of their production processes, on the basis of the knowledge acquired during a period of intense product innovation. It would seem therefore that the arrival of the recession greatly accentuated this phenomenon.

When asked in 1993 about their innovation plans for the next three years, half of the firms said they intended to innovate, whereas only 39 per cent had actually done so in the previous three years. Plans to develop product innovations even outnumbered projects to introduce new processes, marking a definite contrast with the predictions of 1991 when, after a period of intense innovation activity, firms expected to hold such activity at unchanged levels or even sharply reduce product innovation in favour of more process innovation. Having emphasized the unreliability of firms' predictions, we can nevertheless assume that they anticipate an economic recovery and are re -orienting their innovation projects to meet the needs of markets that have regained their dynamism.

Product innovation and process innovation are thus sometimes out of step with each other. Process innovation seems to be less affected by the economic climate, while product innovation is heavily dependent on the dynamics of demand.

Why do firms stop innovating?

These successive surveys of the same population of firms have yielded an interesting performance variable, namely the ability of firms to remain innovative for the duration of the two periods 1986-90 and 1990-92.

Innovation dynamics

The 3 800 or so firms covered by both surveys include firms which were innovative during the first period, and either pursued or ceased their innovation activity in the following period (32.4 per cent and 31.3 per cent respectively). 30.3 per cent of firms said they had developed no new processes or products during either of the periods 1986-90 and 1990-92. Only 6 per cent of firms replied that they had become innovative in the second period.

The inertia of the enterprise population is therefore quite high since nearly two-thirds of firms gave the same reply for both periods. This emphasizes the cumulative nature of innovation. In other words, a firm that is not innovative is more likely to remain that way. However, the chances of remaining innovative when the economic climate deteriorates are much smaller: approximately half the total number of innovative firms stop innovating.

Firms belonging to the traditional sectors of activity are more likely to stop innovating. Consequently, not only do they innovate less, they are also more vulnerable.

By and large, firms which discontinue innovation do so as often for processes as for products. This is particularly true in the case of traditional sectors but also for the basic chemicals and glass industries where the renewal of production processes is a relatively costly operation with the result that cycles tend to be longer. On the other hand, the automobile, parachemicals, pharmaceutical and metal manufacturing industries are more likely to stop product innovation than process innovation. The same applies to aeronautical engineering and electronics.

While small and medium-sized industrial firms (SMIs) innovate less than large firms, they are more sensitive to changing circumstances (Figure 1). Firms with less than 100 employees are able to predict their withdrawal from innovation rather more frequently than larger firms. Firms in traditionally "low technology" sectors also seem to have less difficulty anticipating their withdrawal from the innovation process (Figure 2). They probably have fewer illusions concerning their ability to remain innovative during the next period.

Determinants of withdrawal from the innovation process

The type of innovations developed by a firm over the previous period and the sources it has used to that end are gauges of its ability to remain innovative during both periods. (The typical features and sources of the innovations are therefore placed here on the same level.) The innovation analysis is conducted either globally or separately for processes and products (Annex 1).

◆ Figure 1. ***Percentage of firms ceasing to innovate, by size***
1990-92/1986-90

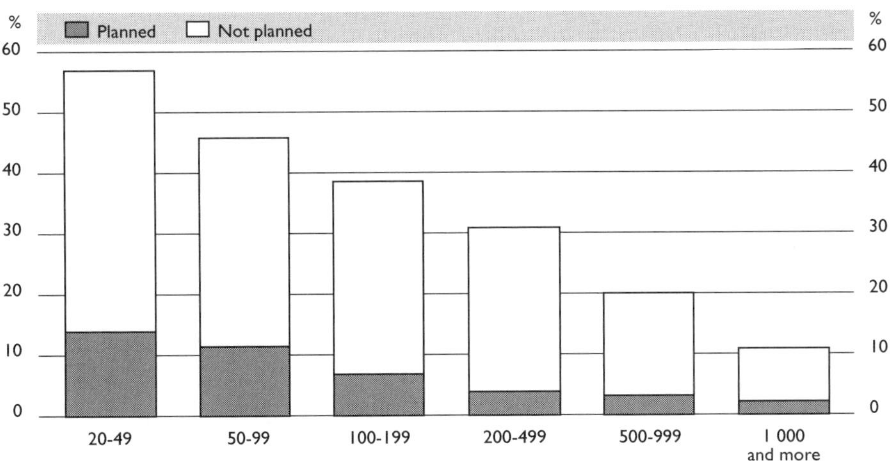

Source: SESSI, 1994.

◆ Figure 2. **Percentage of firms ceasing to innovate, by sector**
1990-92/1986-90

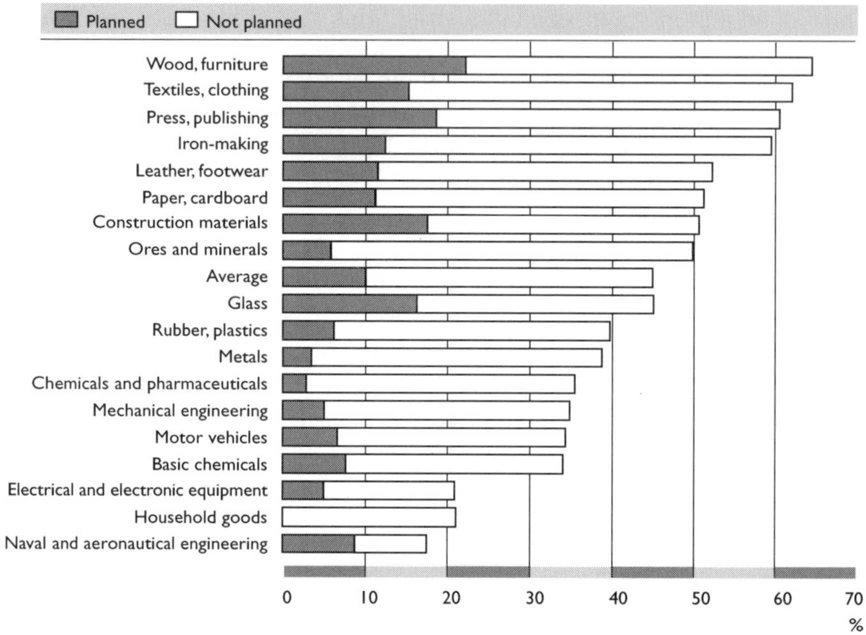

Source: SESSI, 1994.

The propensity to remain innovative basically depends on two factors. First, firm size is a major determinant of ability to continue innovating. As seen earlier, SMIs find it more difficult to remain innovative. This applies to product innovation, while the impact of firm size is smaller (approximately two-thirds) as regards process innovation. The second important factor is the R&D or methods study activity of the firm during the previous period, which influences product innovation continuity to much the same extent as firm size; its impact on process innovation continuity is weaker (by more than two-thirds).

The probability that a firm will cease process innovation is closely related to that of its also ceasing product innovation. The more innovative a firm remains in processes, the greater its chance of continuing product innovation. The reverse is also true.

Three other factors influence a firm's ability to remain innovative, albeit to a lesser extent. The more major innovations (new processes or products) introduced by a firm between 1986 and 1990, the more capable it was of continuing its innovation activity over the period 1990-92. External sourcing or technology, via acquisition of R&D results or licences, is an aid to innovation continuity. Lastly, the greater the share of turnover generated by innovative products, the greater the firm's chances of continuing product innovation during the next period.

Minor but significant impacts are also discernible at firm or sector level:

- At the level of the individual firm, the introduction of new materials and components favours continued innovation activity, particularly product innovation. The introduction of new capital goods does not guarantee that a firm will remain innovative; on the contrary it tends to encourage a cessation of product innovation.
- At sectoral level, a firm operating in a sector where demand pull or technology push tends to remain innovative. Overall, demand has much the same impact as technology, although as regards continued product innovation the determining influence of technology push is twice as strong as that of the opportunities created by demand. The sector to which a firm belongs does not affect the latter's propensity to pursue process innovation. The degree of sectoral concentration has a slightly negative effect on capacity to continue both process and product innovation. Lastly, organisational innovation has no significant impact on the propensity to remain innovative during the next period. The introduction of new processes or products would seem to require further adjustments to the corporate structure.

THE AIMS OF INNOVATION

On average, each of the listed aims is considered more important by large firms than by small firms. The reason for this difference, which is always significant, cannot be stated categorically. One explanation could be that SMIs are presumably less involved in innovative activity at any one time than larger firms and therefore pursue fewer objectives; this would bring the averages down. At the same time it can be argued that innovation activity in large firms is not the same as in smaller ones, which tend to adopt a more passive approach. Since survival is the main driving force for innovation in SMIs, their alternative strategies tend to be more vague.

Over 82 per cent of innovative firms said that they wished to extend their product range (Table 1). This aim comes first followed by the desire to improve quality, which is the most common objective of large firms. The capture of new geographical markets is not a priority for large firms, ranking only eleventh. The innovative activity of large firms is not therefore aimed at extending markets, which

Table 1. **The aims of innovation and their intensity**

Innovating firms	Percentage above 1	All sizes	SMIs		Large firms	
			Average[1]	Ranking[2]	Average	Ranking
To replace products that have become obsolete	61	2.52 (1.44)	2.46 (1.43)	6	3.18 (1.38)	5
To improve the quality of existing products	82	3.34 (1.36)	3.30 (1.37)	2	3.78 (1.04)	1
To extend the product range	83	3.44 (1.38)	3.41 (1.39)	1	3.71 (1.16)	2
To capture new geographical markets	65	2.64 (1.46)	2.63 (1.47)	4	2.77 (1.31)	11
To inject production flexibility	69	2.72 (1.44)	2.68 (1.44)	3	3.26 (1.29)	3
To reduce wage costs	60	2.32 (1.33)	2.26 (1.32)	9	2.98 (1.24)	10
To reduce the consumption of materials	55	2.18 (1.29)	2.11 (1.27)	11	3.00 (1.23)	9
To save energy	44	1.80 (1.11)	1.75 (1.08)	12	2.46 (1.22)	12
To reduce product scrappage	62	2.52 (1.43)	2.46 (1.43)	7	3.20 (1.25)	4
To shorten the product design cycle	60	2.46 (1.44)	2.40 (1.42)	8	3.16 (1.39)	7
To limit environmental damage	54	2.27 (1.40)	2.19 (1.38)	10	3.12 (1.33)	8
To improve working conditions	65	2.61 (1.40)	2.56 (1.40)	5	3.17 (1.23)	6

1. Average intensity given on a Likert scale of 1 to 5.
2. The rankings are the same for the population as a whole. The figures shown in brackets correspond to the standard deviation.

Source: SESSI, 1994.

are already European or world-wide. A central concern would appear to be management of their product range, as borne out by the fact that for firms with more than 1 000 employees the aim of shortening the product design cycle ranks third. Wages and social insurance contributions would not appear to be a particular source of motivation for innovation activity although working conditions are ranked fifth.

By means of a multiple correspondence analysis,[2] it is possible to identify the differences between aims according to firm size and sector of activity.

Firms for which renewal of the product range and the capture of new geographical markets constitute strong or very strong priorities tend to be in such sectors as computers, telecommunications, instruments, electrical machinery and other means of transport (train, aircraft). They are distinct in this respect from firms operating in sectors where product renewal or capture of new markets is very difficult, such as the publishing, metals, leather and footwear, clothing, wood, or paper and cardboard industries. Reduced use of environmentally-damaging technologies, involving areas of activity which generally rely on natural resources and where working conditions are very difficult, is primarily a concern for firms in the chemicals, mining, or refining industries and, to a lesser extent, the rubber and plastics, metals, glass, and paper and cardboard sectors. The reduction of wage costs is an objective mainly for large firms facing keen international competition, as in telecommunications, manufacture of cars, instruments, and electrical machinery, or publishing. Attempts to reduce the consumption of energy and raw materials are most common among firms in the textiles and glass industries, but also concern the chemicals, machinery, paper and cardboard, and rubber and plastics sectors. The aim of shortening design cycles mainly concerns industries where demand is constantly changing (clothing and printing/publishing) and such sectors as other transport and refining, where a global innovation race is under way as competing firms throughout the world seek to develop new models of aircraft and to introduce new fuel processes.

SOURCES OF INNOVATION

In-house R&D remains by far the most important source of innovation. That trade fairs and exhibitions should occupy second place is surprising and contrasts with previous findings (Smith and Vilvei, 1992; Cesaretto and Sirilli, 1992). The relative importance attached to suppliers and customers is compatible with the results of many studies which have stressed the contingent nature of R&D activities for many sectors and sizes of firm (Pavitt, 1984; Archibugi *et al.*, 1991; Napolitano, 1991). Competitors come further down the list, behind other departments of the firm and trade publications. Firms of all sizes consider that, on average, industrial and trade associations and public laboratories do little to spur innovation. Overall, the results of the survey reveal a wide range of external sources in addition to in-house resources.

Table 2 highlights the specific features of large firms relative to SMIs. The results of R&D conducted within the group then assume their full importance, ranking higher than other departments of the firm. The main outside sources are contacts established with suppliers or customers, but they are regarded as less important. General information obtained from databases or trade journals ranks higher than public or private suppliers of R&D results, which concern only a quarter of firms. Links between large firms and public laboratories (university, CEA, CNRS, CNES, for example) are seen as weak, but supplement the in-house network and therefore establish R&D as an essential source of innovation in large firms.

The transfer of S&T knowledge is easier between firms belonging to a group. A group structure is vital for computer firms and, to a lesser extent, for firms in the

Table 2. **Sources of innovation and their intensity**

Innovating firms	Percentage above 1	All sizes	SMIs		Large firms	
			Average[1]	Ranking[2]	Average	Ranking
R&D department	78	3.34	3.28	1	4.09	1
Other departments of the firm	50	2.11	2.05	6	2.74	3
R&D department of the group	32	1.82	1.73	9	2.84	2
Other departments of firms within the group	22	1.43	1.38	14	2.03	11
Universities	24	1.44	1.39	13	1.96	13
Public laboratories	23	1.40	1.35	15	1.96	12
Industrial and trade associations	32	1.61	1.59	10	1.92	14
R&D service providers	25	1.47	1.44	12	1.76	15
Consultancy firms	21	1.36	1.34	16	1.66	16
Materials and component suppliers	55	2.21	2.17	5	2.61	4
Equipment suppliers	58	2.28	2.26	4	2.54	5
Customers	58	2.32	2.30	3	2.53	6
Competitors	44	1.83	1.80	8	2.17	10
Consultation of databanks, patents, models	37	1.64	1.57	11	2.38	9
Conferences, meetings, trade publications	52	2.00	1.96	7	2.50	7
Trade fairs, exhibitions	63	2.34	2.32	2	2.47	8

1. Average intensity given on a Likert scale of 1 to 5.
2. The rankings are the same for the population as a whole. Figures shown in brackets correspond to the standard deviation.

Source: SESSI, 1994.

chemicals, metals and electrical machinery sectors, in contrast to the independence of firms in traditional manufacturing sectors such as clothing, leather, wood and publishing industries.

As before, a factor analysis reveals that some science-based firms tend to use public laboratories (Pavitt, 1984). Most of them operate in the refining, chemicals, telecommunications, instruments and glass sectors. The more traditional sectors, along with the rubber and plastics and metal-working industries, rely more heavily on their component and materials suppliers.

Recourse to consultants or R&D service providers concerns the clothing, telecommunications, publishing and leather industries. However, the nature of such services would seem to differ, according to whether their function is to supplement the work of materials and component suppliers or that of public laboratories. It is also significant that it is mainly SMIs which have recourse to such consultants and other service providers. This explains why the fixed costs associated with R&D and access to new technology are so high for SMIs.

MEANS USED TO ACQUIRE TECHNOLOGY

The means by which firms acquire technology have yet to be defined. The CIS questionnaire proposes 12 possible means, with a distinction in each case between national and international sources.[3] The pattern of technology acquisition has three components: the variety of means used, their degree of complementarity, and frequency of use.

R&D co-operation is the principal means of obtaining scientific knowledge from external sources; it concerns over half the total number of innovating firms and more than two-thirds of large firms (Table 3). The frequency with which firms resort to this means is greater than in the case of other equally intangible means such as contacts with suppliers or customers. Purchase of equipment comes after commissioning of R&D, followed by analysis of competitors' products. Large firms use the different means of acquisition more easily than SMIs. In their case, recruitment is used as a means of acquiring technology just as frequently as the purchase of materials, whereas the latter is more common among SMIs.

The survey shows that these scientific and technological links are mainly established at national level, with international links limited for the most part to EC countries. Furthermore, international links tend on the whole to concern passive forms of knowledge acquisition, such as contacts with customers or suppliers, equipment purchases or analysis of competitors' products.

Two types of sector are clearly identifiable. The first comprises firms which often use external means of acquiring technology, in sectors such as telecommunications, computing, motor vehicles, and other means of transport. To a lesser extent, firms in the rubber and plastics or electrical equipment sectors also rely on their scientific and technical environment. The second type of sector comprises firms in the traditional industries which seem to adopt a more independent approach to developing their technology. According to the results of the survey, the leather, footwear, wood and publishing industries make less use of external sources of knowledge. This sectoral distinction ceases to apply, however, in the case of innovative firms seeking to acquire technology. In this context, computing firms adopt a very different approach, taking advantage of the important exchanges made possible by joint ventures or take-overs of other firms. They also purchase patent rights or organise personnel exchanges but have no more need for R&D links than other firms. Firms operating in the other dynamic sectors mentioned above also tend to have more frequent recourse to joint ventures or the purchase of licences, particularly in the chemicals and refining industries. Traditional sectors such as wood, paper and cardboard, publishing and textiles prioritise information exchanged with suppliers but also with customers. Equipment purchases and recruitment of qualified personnel are used by most firms rather than being specific to any one sector or sectors.

Table 3. **Means of knowledge acquisition, as a percentage of innovative firms**

		France	EC	Europe, excluding EC	United States	Japan	Other	Total
Commissioned R&D	Large firms	55.7	24.0	6.7	16.9	5.2	2.0	67.2
	SMIs	40.6	6.7	0.9	2.0	0.8	0.4	44.6
	Total	41.8	8.1	1.4	3.1	1.2	0.5	46.4
R&D co-operation	Large firms	62.0	31.7	6.6	16.1	5.2	1.5	71.7
	SMIs	44.9	10.9	1.6	1.8	0.6	0.5	51.1
	Total	46.3	12.5	2.0	2.9	1.0	0.6	52.7
Right to use others' inventions	Large firms	18.8	14.9	5.1	15.9	6.2	1.2	35.7
	SMIs	9.9	4.0	0.9	2.0	0.6	0.5	15.4
	Total	10.6	4.8	1.3	3.1	1.1	0.5	17.0
Analysis of competitors' products	Large firms	36.6	43.6	14.1	25.5	24.8	2.9	58.6
	SMIs	27.6	21.3	6.2	8.5	5.7	1.5	38.5
	Total	28.3	23.0	6.8	9.9	7.2	1.6	40.0
Use of consultants	Large firms	15.5	6.9	1.2	2.1	1.3	0.2	20.9
	SMIs	11.6	1.3	0.6	0.5	0.2	0.5	13.4
	Total	11.9	1.7	0.6	0.6	0.3	0.5	14.0
Equipment purchases	Large firms	33.6	29.4	5.5	10.9	5.2	0.7	46.7
	SMIs	28.6	17.3	2.7	2.6	1.9	0.4	42.6
	Total	29.0	18.3	2.9	3.3	2.2	0.5	42.9
Recruitment of qualified personnel	Large firms	40.2	11.0	2.1	5.1	1.6	0.2	43.4
	SMIs	27.0	2.0	0.5	0.4		0.5	29.1
	Total	28.1	2.7	0.6	0.7	0.1	0.5	30.2
Contacts with suppliers	Large firms	54.8	37.9	10.5	14.9	7.0	1.5	61.5
	SMIs	44.0	19.3	3.4	3.1	1.5	0.9	50.2
	Total	44.8	20.7	3.9	4.0	2.0	1.0	51.1
Contacts with customers	Large firms	49.3	35.3	12.4	17.5	6.9	5.5	54.5
	SMIs	46.0	17.1	5.9	4.7	2.3	2.9	49.9
	Total	46.3	18.5	6.4	5.7	2.6	3.1	50.3
Take-over of (all or part of) another firm	Large firms	14.7	10.1	2.0	5.7	0.3	0.2	23.2
	SMIs	8.2	1.3	0.5	0.3		0.4	10.6
	Total	8.7	2.0	0.7	0.8	0.0	0.4	11.6
Joint ventures or partnerships	Large firms	11.2	13.1	2.9	11.5	5.2	1.9	26.0
	SMIs	3.7	2.0	0.6	1.1	0.2	0.7	7.7
	Total	4.3	2.9	0.8	1.9	0.6	0.8	9.1
Staff exchanges	Large firms	8.0	7.6	1.2	5.9	1.0	0.0	16.5
	SMIs	3.2	1.0	0.1	0.2	0.1	0.3	5.4
	Total	3.6	1.5	0.2	0.6	0.1	0.3	6.2

Source: SESSI, 1994.

Complementarity between the means used can be studied by means of a partial correlation matrix (*i.e.* once firm size in terms of turnover has been taken into account; Table 4). Innovating firms are seen to present a variety of linkage patterns: often enough, two firms will have the same number but not the same types of links. The matrix shows that the different means used by innovating firms are complementary (see also, Arora and Gambardella, 1990). The degree of

Table 4. **Matrix of correlation (taking account of firm size) between the different means of knowledge acquisition**[1]

		Q1	Q2	Q3	Q4	Q5	Q6	Q7	Q8	Q9	Q10	Q11
Commissioned R&D	Q1	1.00										
R&D co-operation	Q2	0.21	1.00									
Right to use others' inventions	Q3	0.15	0.12	1.00								
Analysis of competitors' products	Q4	0.23	0.13	0.16	1.00							
Use of consultants	Q5	0.17	0.12	0.19	0.20	1.00						
Equipment purchases	Q6	0.08	0.09	0.06	0.18	0.17	1.00					
Recruitment of qualified personnel	Q7	0.18	0.12	0.18	0.24	0.20	0.26	1.00				
Contacts with suppliers	Q8	0.18	0.18	0.05	0.33	0.17	0.28	0.29	1.00			
Contacts with customers	Q9	0.16	0.14	0.10	0.32	0.12	0.20	0.29	0.42	1.00		
Take-over of (all or part of) another firm	Q10	0.13	0.07	0.17	0.17	0.15	0.07	0.19	0.12	0.14	1.00	
Joint ventures, partnerships	Q11	0.17	0.13	0.23	0.16	0.14	0.08	0.22	0.13	0.14	0.29	1.00
Staff exchanges	Q12	0.12	0.09	0.17	0,13	0.18	0.11	0.20	0.13	0.12	0.23	0.34

1. Level of significance of coefficients = 1 per cent.

Source: SESSI, 1994.

complementarity is often significant, but in many cases low. Thus only a few configurations can be considered:

- Personnel exchanges, joint ventures and take-overs of part or all of a firm are relatively highly correlated. The same firm consistently use these radical means of acquiring technology;
- The same applies to contacts established with suppliers and customers. Vertical linkage both upstream and downstream is thus the usual case;
- Vertical links are often accompanied by analysis of competitors' products and recruitment of qualified personnel;
- Recruitment of qualified personnel is accompanied in the case of many firms by information links with suppliers, but also by equipment purchases;
- The linkage between commissioned R&D and R&D co-operation is relatively weak, suggesting the need for separate analysis.

Although firms frequently use external sources, the pattern of use differs widely across firms. An alternative approach to this phenomenon is simply to consider the range of means used by innovative firms to acquire external technology. This range can be measured on a scale of 0 to 12 representing the different means employed simultaneously during the period under review. Only a few firms stated that they had not used any means of acquiring S&T knowledge, whereas approximately half used at least four different means. Less than 5 per cent used more than nine out of the 12 means proposed.

On average, SMIs use more than three different means, compared to over five in the case of firms with more than 1 000 employees. The range of means used also differs across sectors. Figure 3 below shows that there are limits to the innovation gap between the most and the least innovative sectors. For example, firms specialising in such sectors as computers, precision instruments and chemicals/pharmaceuticals tend not to use a wide variety of means to acquire technologies. By contrast, the sectors frequently regarded as "low-technology sectors", such as textiles and furniture, are very active in their use of outside sources. The diversity of means used by these innovative firms is a clear indication that R&D is not the central component of technology acquisition. The automobile industry is the sector with the widest range of acquisition strategies. Contrary to what the R&D data suggest, this sector appears very dynamic technologically. Links with suppliers, researchers and information sources are juxtaposed here to form a particularly wide-ranging network to serve innovators.

This indicator of variety is interesting in that it establishes a quantitative ranking of the sectors that receive and transmit knowledge, independently of the technology flow matrices customarily used by econometricians.

◆ Figure 3. ***Variety of means used to obtain knowledge, by sector***

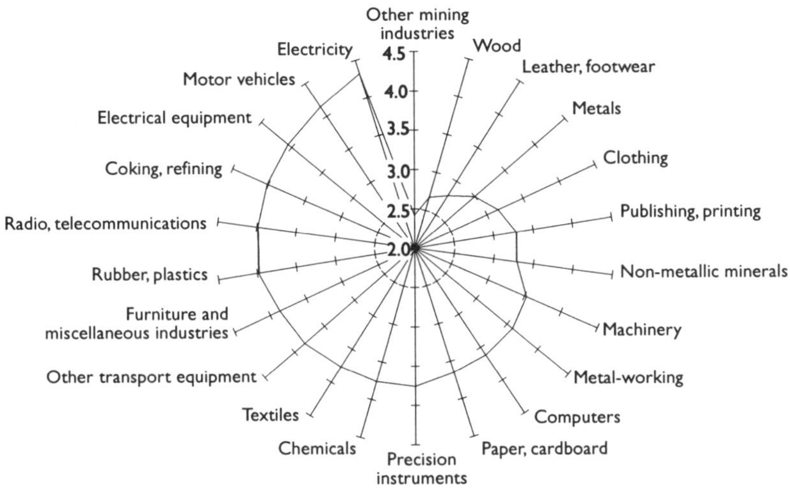

Source: SESSI, 1994.

EXTERNAL ORGANISATION OF R&D ACTIVITIES

In the section that follows innovation networks will be considered from the standpoint of R&D links. An attempt will be made to identify more clearly those means of knowledge acquisition which are complements to or substitutes for R&D links, by incorporating the theoretical problems specific to R&D.

Frequency, variety and complementarity of means of knowledge acquisition

As seen earlier, R&D links are extremely frequent. Cross-sector analysis shows the importance of the industrial structure, since the machinery and metal-working sectors account for over 30 per cent of R&D co-operation, in spite of the small number of innovative firms which use this means of acquiring technologies. The chemicals industry comes before electrical equipment and precision instruments (over 7 per cent each). Computers and telecommunications account for only 4.3 per cent of co-operation agreements. Table 5 shows that the top four sectors account for approximately half of the total R&D links in manufacturing industry. This cross-sector distribution is noticeably different from those found by other studies (see Hagedoorn and Schakenraad, 1991). This is not surprising since an affirmative answer from firms can cover many individual agreements. Furthermore, the data

Table 5. **Distribution of R&D links in manufacturing industry**

Sectors	Percentage of innovating firms which have R&D links	Sectoral distribution		Sectors	Percentage of innovating firms which have R&D links	Sectoral distribution	
		%	Cumulative %			%	Cumulative %
Machinery	71.0	16.4	16.4	Radio, telecommunications	84.7	3.7	86.5
Metal-working	68.2	15.0	31.4	Metals	73.2	2.4	91.3
Chemicals	80.5	8.7	40.1	Paper, cardboard	63.3	2.4	88.9
Rubber, plastics	71.1	7.4	47.5	Other transport equipment	79.8	2.3	93.6
Precision instruments	77.0	7.3	54.8	Wood	55.3	1.6	95.3
Electrical appliances	77.2	6.1	60.9	Clothing	37.6	1.2	96.5
Textiles	57.7	4.9	65.8	Leather, footwear	48.5	1.0	97.5
Publishing, printing	49.0	4.7	70.5	Other mining industries	78.8	0.9	98.5
Non-metallic minerals	64.6	4.4	74.9	Coking, refining	91.0	0.7	99.2
Furniture and miscellaneous industries	66.3	4.0	78.9	Computers	54.7	0.6	99.7
Motor vehicles	92.3	3.9	82.8	Electricity	82.4	0.3	100.0

Source: SESSI, 1994.

collection methods are not comparable. Lastly, R&D agreements are only one example of the various agreements concluded between establishments.

Vertical links dominate, whereas R&D links with competitors, whether in the form of subcontracting or co-operation agreements, are rare (representing less than 6 per cent). However, competitors represent a source of innovation for more than 44 per cent of firms.

Table 6 also shows that external R&D links are established mainly with materials and equipment suppliers. The importance of co-operation is borne out by the fact that links with customers rank third. Industrial and trade associations and other firms in the same group also feature very largely in exchanges of R&D results, well ahead of universities, which even so have links with a quarter of the innovative firms which commission R&D or engage in R&D co-operation. Although materials suppliers continue to occupy first place as the partners of large firms, more than 55 per cent of such firms use the services of the group's central laboratory or universities.

Table 6. **Percentage of firms which commission R&D or engage in R&D co-operation**

	All sizes	Ranking ↙↘	SMIs	Large firms	Ranking
Materials and component suppliers	55.2	1	54.8	59.0	1
Equipment suppliers	54.0	2	54.0	54.4	5
Customers	43.7	3	44.2	38.5	8
Industrial and trade associations	35.1	4	33.3	51.2	6
Other firms in the group	27.9	5	24.8	55.0	4
Universities	25.9	6	22.6	55,.3	3
Group's central laboratory	24.2	7	20.7	55.7	2
Public laboratories	21.5	8	19.2	42.0	7
Consultants	19.3	9	18.7	24.6	9
Competitors	8.5	10	7.9	13.9	10

Source: SESSI, 1994.

Practices differ considerably across sectors. Firms which rely on co-operation or subcontracting for their R&D establish these links with either suppliers or universities. They thus obtain one of two types of knowledge: technological knowledge, which essentially enables them to adapt capital goods and materials or, equally, products, or scientific knowledge obtain from public laboratories. Firms in the sectors of publishing, textiles, rubber and plastics, and also motor vehicles and furniture, all belong to the first category. The second category comprises firms in the chemicals, refining, instruments and wood industries.

Some firms organise most of their R&D activities with other members of their group. This is particularly the case with large firms which also have links with public laboratories. Other firms, however, in the leather, clothing and wood industries, prefer to use consultants. Firms in the refining and computing sectors use other firms in the group, although computing is a special case since computer firms also establish R&D links with their competitors.

The internationalisation of R&D, identified here as the propensity to establish R&D links with foreign partners, only concerns firms belonging to groups and sectors such as the refining or chemicals industries.

The matrix of partial correlations between different types of partners (Table 7), calculated for all innovative firms, shows complementarity between R&D linkages established with the different partners (at a 5 per cent level of significance). Although R&D links are frequent, the low correlation coefficients suggest that the configurations adopted by firms are rather volatile. The only links that stand out are those which are similar, such as intra-group links (Questions 7.1 and 7.2), links with suppliers (Questions 7.4 and 7.5) and links with public laboratories (Questions 7.9 and 7.10). They reflect a focusing of R&D links on these types of partners, but also a degree of confusion between the items proposed, which consequently seem too precise. The only notable correlation is ultimately the one between links with suppliers and links with customers, which underscores the importance of vertical linkages in R&D.

With these findings it is once again possible to analyse the variety of R&D partners, which increases with the size of firms. As with the diversity of means used to acquire technology, the diversity of partnerships established in order to obtain R&D results mainly concerns firms with more than 500 employees and in particular very large innovative firms.

At sectoral level, the computing and refining industries, together with metals, paper and cardboard, leather and furniture, obtain their R&D results from several different sources. The telecommunications, precision instruments and chemicals/pharmaceuticals sectors use a much narrower range of knowledge sources (Figure 4).

This second graphic ranking of sectors shows that certain high-tech sectors such as telecommunications and chemicals have only a few scientific and technological partners. More traditional sectors such as metal products or paper and board need to assemble a wider range of R&D skills. The automotive industry, which was seen earlier to channel more technologies than any other, is supplanted here by sectors that are more science-based, such as computing or aviation.

Table 7. **Matrix of partial correlations (after allowance for firm size) between different types of R&D partners**[1]

		Q1	Q2	Q3	Q4	Q5	Q6	Q7	Q8	Q9
Group's central laboratory	Q1	1.00								
Other firms in the group	Q2	0.37	1.00							
Customers	Q3	0.11	0.20	1.00						
Equipment suppliers	Q4	0.05	0.15	0.35	1.00					
Materials and component suppliers	Q5	0.15	0.19	0.39	0.52	1.00				
Competitors	Q6	0.06	0.12	0.18	0.19	0.17	1.00			
Consultants	Q7	0.02	0.05	0.10	0.15	0.15	0.08	1.00		
Industrial and trade associations	Q8	0.12	0.08	0.22	0.21	0.23	0.10	0.20	1.00	
Universities	Q9	0.16	0.11	0.17	0.08	0.09	0.07	0.14	0.26	1.00
Public laboratories	Q10	0.13	0.10	0.17	0.11	0.13	0.07	0.13	0.24	0.43

1. Level of significance of coefficients = 1 per cent.
Source: SESSI, 1994.

◆ Figure 4. ***Variety of R&D partners, by sector***

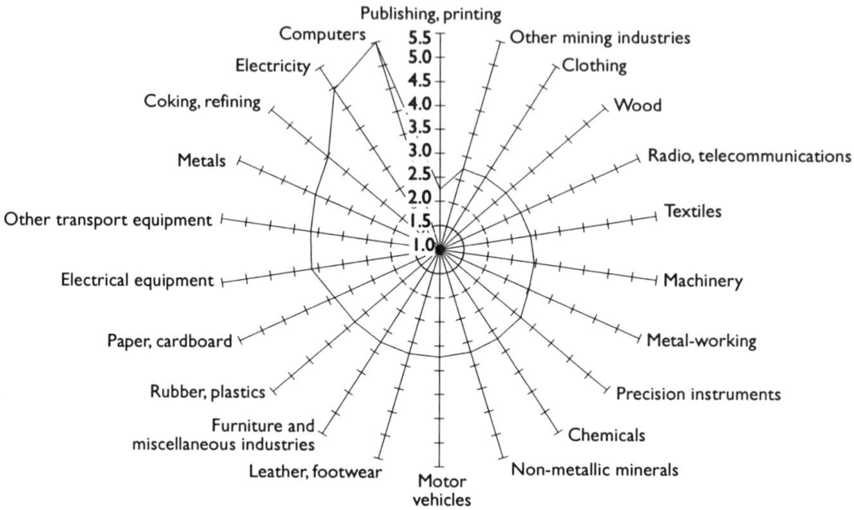

Source: SESSI, 1994.

Determinants of R&D linkage

The links established by firms in their R&D activity thus contribute significantly to their absorption of S&T knowledge. Cohen and Levinthal (1989), in a groundbreaking article, suggest that absorption of R&D results is a function of firms' internal expenditure on R&D. The innovation survey makes it possible to test the premise that if a firm performs in-house R&D this increases the likelihood that it will establish R&D links outside. However, major (sectoral) disparities show that firms also use other means of absorbing new technologies. In other words, their capacity to absorb technology does not depend solely on their in-house R&D performance. Consequently, the hypothesis put forward by Cohen and Levinthal is tested on the different means used by firms to acquire R&D results, once their other strategies for technology acquisition have been taken into account.[4]

After allowance for the effects of firm size and sector, analysis of competitors' products has a significant impact on the probability of R&D being commissioned. (The findings are given in Annex 2.) Performance of R&D (dummy variably here) increases the probability of R&D links being established. The distinction made between research according to the Frascati definition and more informal research[5] (which is not included in the survey conducted by the French Ministry of Higher

Education and Research) does not appear to be crucial here since the effects are similar. However, the influence of R&D is relatively weak. On the other hand, consultancy services and contacts with suppliers seem to be twice as influential. Use of general information is also significantly correlated. However, such complementarity is not significant in the case of such means as the right to use others' inventions, analysis of competitors' products, recruitment of qualified personnel or equipment purchases. Use of databases, patents or models is generally more influential than recourse to R&D and almost twice as influential as the use of conferences or trade journals.

The greater the extent to which knowledge can be classified within a particular sector, the more likely firms are to subcontract their research activities. The fact of belonging to an innovative sector where innovation is spurred by technology push or keen competition does not mean that firms are significantly more likely to establish R&D links. However, the effects of belonging to a particular sector are of the same magnitude as those of performing R&D.

A distinction can be made between unilateral and bilateral R&D links. The fact of performing R&D ceases to be determinant when it is a case of explaining the propensity to commission R&D or engage in R&D co-operation. On the other hand, the right to use others' inventions and joint ventures become significant. The differences between the factors which determine whether a firm resorts to co-operation or subcontracting are particularly noticeable at the level of the coefficients. For example, commissioning of R&D concerns above all firms where knowledge is more codified or firms which use databases. Similarly, the purchase of licenses is more characteristic of firms which commission R&D. Analysis of competitors' products is also more common among firms which prefer R&D subcontracting arrangements. Unlike firms which favour collaborative R&D, however, they establish no contacts with their customers.

By repeating the analysis for each R&D partner, it is possible to show that in-house R&D and databases are important only for firms which wish to establish R&D links with universities or public laboratories.[6] R&D on the Frascati definition seems slightly more important for links with customers. The classification of knowledge would seem to be just as influential as performance of R&D in the acquisition of S&T knowledge from public laboratories.

Knowledge acquisition is a process which necessitates spending on R&D in certain cases only. The findings given here indicate very broadly that Cohen and Levinthal (1989) were right to relate acquisition to R&D spending and to the characteristics of the knowledge to be acquired. However, this view would seem to apply exclusively to high technology links, *i.e.*, with public laboratories and industrial or trade associations.[7] This is particularly the case with knowledge acquired from suppliers or from other members of the group, which seems to depend respectively on recruitment of qualified personnel and staff exchanges. This absence of in-house R&D is also apparent in the case of R&D commissioning, which would disprove the idea of straightforward complementarity between internal and external R&D. Thus, to acquire R&D results, firms make two kinds of choice: they choose among the different means of acquisition according to the characteristics of the knowledge sought;[8] and, in the more conventional view, they make a trade-off between in-house and external R&D (Friedman *et al.*, 1978; Teece, 1988; Riedle, 1989). In both cases, external knowledge acquisition requires that additional resources be made available to facilitate or accompany the flow of R&D results.

Motivation

The foregoing analysis of the determinants of R&D linkage has attempted to identify the constants associated with R&D co-operation or commissioning. However, it is difficult to infer from these what actually motivates the firms concerned. The French survey has introduced a question on this subject.[9]

Collaborative or commissioned R&D can be motivated, first, by the need to obtain results which the firm is incapable of producing itself. Second, even if a firm possesses the competencies for carrying out an R&D project, it may not have the resources needed in order to do so unaided. Thirdly, by reducing additional workload firms are able to hold costs down. A fourth feature of inter-firm links is their duration. For example, one question asks whether the firm may have a case-specific need for R&D in order to deal with particular problems (in spite of the uncertainty and the specific character of the resources assigned, this question is clearly linked to the transaction cost approach).

Several more specific questions have been added to these traditional questions concerning firms' motives for establishing R&D links:

For example, R&D linkage also contributes to technology monitoring, since a firm can become familiar with the different technological options offered by their various partners.

When a firm in a group exchanges R&D results with the group's central laboratory, its motives for doing so do not always correspond to a wish of its own or to financial flows. The reply "Group policy" was designed to accommodate such cases.

The French survey also introduced three more specific questions concerning corporate strategies:

- The deterrent factor: external R&D linkage can help firms to structure their environment. R&D agreements may be more conspicuous than in-house R&D, and may consequently act as a more of a deterrent to competitors. The deterrent will be all the more credible if the resources acquired are exclusive. From this standpoint R&D links may be regarded as entry barriers (see DeBresson *et al.*, 1991).

– On the other hand, the fact of establishing external R&D links or, better still, using an affiliate for R&D purposes, may be seen as a way of introducing competition. Exposing internal research to competition can motivate firms (in the traditional context of the innovation race), limit their opportunism [in the context of "second sourcing" (Farrel and Gallini, 1988)] or minimise the risk of technological lock-in by keeping different options open.
– Lastly, outside sourcing of R&D may also enable firms to assess R&D costs more accurately. The idea, in the management context, is to find out whether it might not be in firms' interests, in view of management difficulties (loss of control over the R&D function) or tax incentives, to increase the number of contracts with outside bodies.

Two-thirds of firms commission R&D or engage in collaborative research for reasons of technological complementarity (Table 8), which falls into three categories: technological complementarity between the skills of the different firms; case-specific complementarity, given that research is often conducted in response to specific technological problems; and, lastly, complementarity of scale, since the size of certain projects hampers investment. R&D partnerships also provide an opportunity for reducing R&D costs and for monitoring technology. The need for basic R&D is mentioned by more than a third of large firms but is not a major source of motivation for smaller firms.

Case-specific complementarity is a motivation for firms in the metals, wood, refining, chemicals and other transport sectors, which wish to obtain basic R&D results. The same objective is to be found in the telecommunications sector where, however, technological complementarity would seem to be more important.

Table 8. **Motives for commissioned or collaborative R&D**

	All sizes	Ranking ↙↘	SMIs	Large firms	Ranking
Obtain additional R&D results	65.8	1	65.1	72.2	1
Case-specific R&D needs	60	2	59.5	65.2	3
Execute R&D projects which the firm would have been unable to carry out unaided	57.4	3	56.4	66.4	2
Technology monitoring	40.5	4	39.5	49.3	4
Benefit from R&D results at lower cost	39.7	5	39.1	44.8	5
Group policy	21.2	6	20.1	30.4	7
Deter potential competitors	18.6	7	19.4	11.6	8
Basic R&D requirements	18.3	8	16.3	35.8	6
Obtain a better assessment of R&D costs	14.8	9	15.5	9.1	9
Expose in-house R&D to competition	9.9	10	10	8.8	10

Source: SESSI, 1994.

Complementarity of scale is a characteristic of the high-technology sectors where R&D spending reaches high levels. On the other hand, complementarity of scale and basic research are rare in the textiles, publishing and furniture sectors.

Firms in the clothing industry, in particular, but also those producing electrical equipment, furniture (and miscellaneous industries), leather and chemicals/pharmaceuticals experience more difficulty managing R&D than other firms. The opportunity for technology monitoring is mentioned more frequently by firms in the wood, refining, glass, metals, rubber/plastics and textiles industries.

Lastly, the aim of deterring potential competitors is very strong in the automotive industry where R&D links provide subcontractors or equipment suppliers with an opportunity to "lock in" their relationship with their principals who, given the increase in specifically assigned resources, would find multiple contracting difficult.

DIFFUSION OF PRODUCT INNOVATIONS

These R&D links having been identified, albeit only roughly, the next step is to try to assess their impact. This brings one up against the traditional problems of measuring output. The share of total turnover generated by innovations gives some idea of the impact of product innovation on firms' performance. Overall, less than 3 per cent of firms have an innovation-generated turnover exceeding 70 per cent of total turnover (see Table 9). Less than 25 per cent of innovating firms report a product innovation share exceeding 30 per cent of turnover. For more than a third of innovative firms, product innovations represent between 10 and 30 per cent of their turnover. SMIs obtain only 9 per cent of their turnover from product innovations. The rate is 20 per cent for large firms, with the result that 17 per cent of the turnover achieved by French manufacturing industry as a whole comes from product innovations introduced less than three years ago. Cross-sector analysis shows that product range renewal is much more frequent in the computer industry (Figure 5). On the other hand, product innovation in the traditional sectors of industry rarely accounts for more than 70 per cent of turnover.

There is no significant link between firm size[10] and the share of innovation-generated turnover, contrary to the results of the previous survey where product innovation accounted for a larger share of sales and exports in the case of large

Table 9. **Innovation-generated shares of turnover and exports**

	0-10%	10-30%	30-70%	70-100%
Percentage of innovation-generated turnover	41.1	39.5	16.8	2.5
Percentage of innovation-generated exports	58.4	24.5	12.6	4.5

Source: SESSI, 1994.

◆ Figure 5. **Innovation-generated turnover shares by sector**

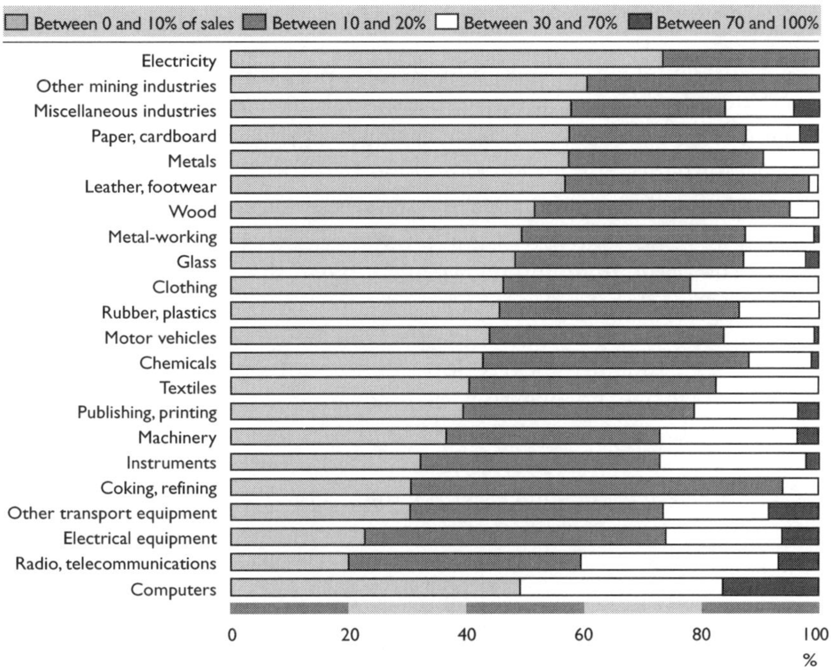

Source: SESSI, 1994.

firms. This result, which is consistent with those obtained for the Netherlands (see Brouwer and Kleinknecht, 1994), is indicative of the important change in the population of innovative firms, whose numbers have declined. Their innovation activity is more intense than during the previous period, because the recession has weeded out the least innovative firms. At the aggregated level this variable should therefore be treated with caution, particularly when the same population is being monitored.

CONCLUSION

The CIS has enabled us to make headway on several aspects of the innovation process which had been unclear. In particular, we feel it has given a truly statistical dimension to concepts often originating in monographs or based on limited samples. The innovation surveys thus provide a good description of the economic and technological processes at work in industry as a whole. They are a tool for the advancement of knowledge and a valuable aid for the design of public policies and for corroborating or guiding research efforts.

Survey data are being made available to university research teams requesting them, which guarantees a growing number of inputs for the future. In addition, the fact of matching the survey data with data from other administrative surveys or tax data available in France opens up many avenues of research which will complement the work on international comparisons done by EUROSTAT.

The work carried out in France in connection with the CIS questionnaire has also provided an opportunity to obtain a better grasp of the difficulties of this kind of approach. The preliminary interviews encouraged the France's Ministry of Higher Education and Research to initiated a special study on innovation costs. Thus, a monograph on some 20 enterprises (carried out by BETA of Strasbourg University) should enable us to identify or develop the accounting categories necessary for a statistical approach to the question of innovation costs. At the same time, the weight of the service sectors in French GDP means that they should be included in future surveys. The specific characteristics of innovative processes at work in service firms therefore need to be analysed so that satisfactory questionnaires can be drawn up.

At the same time, a review of the CIS questionnaire as submitted to enterprises highlighted the questions and wording which caused problems. We have indicated in the text the difficulties encountered regarding such terms as technology monitoring, but we can also enter reservations on, *inter alia*, the aims of innovation insofar as the items worded positively were more frequently selected by firms which have difficulty in perceiving positively an act (to innovate) or a reduction, fall or decrease, but find it easier to accept an improvement or an expansion. These difficulties coupled with the crucial issue of translation should therefore be dealt with before the next innovation survey is launched in 1997.

Annex 1

VARIABLES

The explanatory variables were constructed from replies to the innovation survey carried out in France in 1991:

- the size variable used in the econometric calculations was turnover; we used the logarithm of this variable (the square of the variable was also taken);
- the nature of technological innovations was covered by five yes/no (1 or 0) questions: product innovations (U1, product improvement; U2, introduction of a product new to the market; U3, introduction of a product new to the firm) and process innovations (U4, introduction of technology new to the firm; U5, substantial process improvements). In this way an indicator at three levels (0, 1, 2) was constructed such that the Radical = U2 + U4;
- commercial innovations;
- innovations in packaging methods;
- organisational innovations.

Firms were asked to assess the importance of the three following sources of innovation on the basis of a Likert scale of 0-3:

- R&D: impact of in-house R&D work;
- methods: importance of the technical studies department, etc.

The arithmetic mean of these last two replies determines the R&D-method variable here:

- materials: the role played by the introduction of new materials or components;
- introduction of new equipment;
- patents owned by firms;
- R&D results acquired outside the firm;
- licences purchased or patents acquired or sold.

The mean of the last two variables gives the variable "external R&D or patent acquired", used in this analysis.

Two other variables were used:

- share of innovation-generated turnover: firms were asked to place the percentage share of turnover generated by product or process innovations in one of the following categories: 0-10, 10-30, 30-70 or 70-100 per cent;
- sectoral indicators calculated at the two-digit level of the classification system.
- the sectoral indicator showing the influence of demand pressure was constructed from the mean of replies, using 4 degrees of the associated Likert scale (none, slight, moderate, substantial) to the question, "From the firm's point of view, do you consider innovation to be determined by market pull?";

- the same method was used to determine the TECHNO variable which was obtained from answers to the following question: "From the firm's point of view, do you consider innovation to be determined by technology push?";
- the annual survey of firms by the Ministry for Industry provided the Hirschmann index (H) as a measure of market concentration calculated from sales for 1988, at level 100 of the French nomenclature of activities and products (NAP).[11]

THE RESULTS

Annex Table 1.1. **Descriptive statistics of the independent variables**

N = 2 555		Mean	Standard deviation
Characteristics of innovations over the period 1986-90	Radical innovation	0.77	0.77
	Commercial innovation	0.19	0.39
	Packaging innovation	0.22	0.41
	New equipment	1.58	1.10
	New materials or components	1.22	1.11
Sources of innovation used over the period 1986-90	R&D or methods studies	1.52	0.97
	R&D or methods studies × group R&D	1.28	2.39
	External R&D or acquired patents	0.44	1.36
	Patents	0.61	0.98
	Size (log)	17.9	1.62
Other characteristics	Share of innovation-generated turnover[1]	0.72	0.86
	Cessation of product innovation[1]	0.44	0.50
	Cessation of process innovation[1]	0.44	0.50
Sectoral indicators	Concentration (log)	2.39	0.62

1. These values should normally be calculated on the reduced populations (2 230 or 1 981), though differences are small.

Source: SESSI, 1994.

Annex Table 1.2. **Determinants of withdrawal from innovation** [1, 2]

	Model	Probit		
	N	2 555	2 230	1 981
		Explained variable		
	Explanatory variables	Innovation	Production innovation	Process innovation
Characteristics of innovations over the period 1986-90	Radical innovation	–0.070 (0.015)	–0.087 (0.017)	–0.041 (0.016)
	Packaging innovation	–	–	–0.069 (0.028)
	New equipment	–	0.040 (0.012)	–0.041 (0.012)
	New materials or components	–0.021 (0.01)	–0.052 (0.012)	–
Sources of innovation used over the period 1986-90	R&D or methods studies	–0.127 (0.013)	–0.124 (0.016)	–0.074 (0.014)
	R&D or methods studies × group R&D	0.026 (0.005)	0.021 (0.006)	–
	External R&D or acquired patents	–0.045 (0.017)	–0.060 (0.019)	–
	Patents	–0.037 (0.012)	–0.033 (0.013)	–
Other characteristics	Size (log)	–0.079 (0.008)	–0.079 (0.009)	–0.049 (0.008)
	Share of innovation-generated turnover[1]	(0.015)	–0.068	
	Cessation of product innovation[1]			0.357 (0.025)
	Cessation of process innovation[1]		0.349 (0.025)	
Sectoral indicators	Technology[3]	–0.050 (0.012)	–0.063 (0.014)	–
	Demand[3]	–0.050 (0.011)	–0.038 (0.013)	–
	Concentration (log)	–0.041 (0.020)	–0.048 (0.023)	–
	Constant	1.722 (0.145)	0.010 (0.365)	1.055 (0.142)
	Maximum likelihood	–1 447	–1 107	–1 119
	McFadden R^2	0.174	0.280	0.175

1. The coefficients are the marginal effects. They show the effect of a change in the variable on the probability of remaining innovative. The estimates are made about the mean. The standard deviation is shown in brackets. The coefficients are all significant at the 5 per cent level. The non-significant variables (10 per cent level) are dropped. The values are then re-estimated.
2. The value of the coefficients is not easy to interpret, unlike their sign. However, for the non-standardised variables the coefficient can be multiplied by the variable's standard deviation (see Table 4). In this way, we obtain the magnitude of the effect when the standard deviation of the variable changes. Thus, the R&D effect is (0.127 × 0.97) whereas the concentration effect is around five times smaller (–0.039 × 0.62).
3. Standardised variables.

Source: SESSI, 1994.

Annex Table 2.1. **Characteristics of R&D sub-contracting and co-operation: probit model results**[1, 2]

	R&D N = 1 542	R&D link	R&D sub-contracting	R&D co-operation
Individual caracteristics	Turnover (log)	0.046 (5.727)	0.034 (3.794)	0.038 (4.426)
R&D	R&D (Frascati definition)[3]	0.087 (1.713)	–[5]	–
	R&D (Oslo definition)[3]	0.066 (2.198)	–	–
Alternative means of acquisition	The right to use others' inventions	–	0.115 (3.071)	0.066 (1.797)
	Competitors' products[3]	–	0.091 (3.017)	–
	Consultancy services[3]	0.173 (4.29)	0.167 (4.016)	0.102 (2.539)
	Contacts with suppliers[3]	0.114 (4.373)	0.099 (3.464)	0.099 (3.348)
	Contacts with clients[3]	0.080 (3.124)	–	0.061 (2.112)
	Joint ventures or partnerships[3]	–	0.142 (2.796)	0.175 (3.467)
General sources of information	Databases, patents, models	0.044 (2.901)	0.058 (3.888)	0.026 (2.137)
	Trade conferences, trade press	0.024 (1.994)	–	–
Sectoral indicators	Classification[4]	0.120 (2.532)	0.244 (4.833)	0.079 (1.608)
	Demand pull	0.176 (2.464)		–
	Log likelihood	–787.04	–942.44	–980.83
	χ	271.76	252	32.20
	D.L.	10	8	4

1. See Annex 3 for the definition of variables.
2. The coefficients are significant at the 95 per cent level. All the regressions include a constant. The elasticities are the marginal effects calculated about the mean. The non-significant variables are: R&D on the Frascati definition,[3] R&D on the Oslo definition,[3] innovation,[4] demand,[4] technology,[4] concentration, fairs/exhibitions, equipment purchase, hiring of qualified personnel, personnel exchanges.
3. Dummies.
4. Sectoral indicators.
5. – = Non-significant coefficient.

Source: SESSI, 1994.

Annex Table 2.2. **Explanatory variables: mean and standard deviation**

	Variables	Mean	Standard deviation
Characteristics	R&D, Frascati definition	0.10	0.30
	R&D, Oslo definition	0.72	0.45
Alternative means of acquisition	The right to use others' inventions	0.19	0.39
	Competitors' products	0.43	0.49
	Consultancy services	0.14	0.35
	Recruitment of qualified personnel	0.32	0.46
	Contacts with suppliers	0.52	0.5
	Contacts with clients	0.51	0.5
	Joint ventures or partnerships	0.11	0.31
	Exchange of personnel	0.06	0.25
General sources of acquisition	Databases, patents, models	1.76	1.04
	Trade conferences, trade press	2.08	1.14
	Fairs/exhibitions	2.35	1.2
Sectoral indicators	Classification	1.34	0.27
	Demand	1.72	0.17
	Technology	2.37	0.1

Source: SESSI, 1994.

Annex 3

LIST OF DETERMINANTS OF THE PROPENSITY TO HAVE R&D LINKS

All the questions on R&D partners are binary (0–1).

– The size variable used here is turnover; the centred logarithm of the variable is taken here (the square of the variable is also taken to verify whether the relationship is linear).

– The export ratio is the ratio of exports to turnover for 1990.

The individual indicator variables (0 or 1).

– R&D-Frascati: firms which carry out R&D (at least one full-time researcher or equivalent) and which took part in the R&D survey carried out by the Ministry for Research between 1988 and 1991.

– R&D-Oslo: enterprises which state that they perform R&D but which do not figure in the Ministry for Research survey.

These last two variables are mutually exclusive.

– The alternative means of acquiring technology correspond to the items of the question and are shown in Table 8. All the questions are considered on a yes-no basis without specification of the international or national nature of the means employed. Only the take-over of all or part of another firm has not been taken into consideration. Where the correlations are too evident, the variable has been dropped. For example, where a firm states that it has contacts with suppliers, the purchase of equipment has not been introduced into the model.

– Three sectoral indicators were calculated at the two-digit level of the NACE (General Industrial Classification of Economic Activities within the European Communities):

– the "demand pull" variable reflects the role of the market in the technological innovation process in a given sector;

– the "technology push" variable reflects the role of technological knowledge production in the innovation process in a given sector;

– the "classification" variable corresponds to the average importance accorded in a given sector to data banks, patents, models, etc. As a first first approximation, it contrasts with the tacit nature of certain technologies.

NOTES

1. The author would like to thank the staff of SESSI for the technical support provided for the survey, particularly J.P. François and T. Lehoucq, and also all members of the statistics department at the French Ministry of Higher Education and Research who provided very valuable practical and technical support.

2. The factorial designs have not been reproduced here due to lack of space and because they are difficult to read.

3. The questionnaire equally allowed for the possibility of outward transfers of knowledge by firms. This point will not be pursued as very few firms replied to the relevant question.

4. We have only tested the relationship between external sources of knowledge and the presence or not of an absorption capacity. We cannot say anything about the growth of this capacity.

5. See also OECD (1992), "OECD Proposed Guidelines for Collecting and Interpreting Technological Innovation Data – Oslo Manual", OCDE/GD(92)26, Paris, p. 47.

6. This result is consistent with that obtained by Nelson and Wolff (1995), and, more generally, with other studies based on Yale data.

7. This result may also reflect a bias of the Yale I survey in favour of large firms and certain science-based sectors.

8. We have included only one sectoral indicator here. Data obtained from surveys on firms' R&D (internal R&D expenditure, share of basic R&D, researchers' formal qualifications, number of support staff per researcher) and from the European patent database (variety of technologies mastered by a firm) are currently being used to make progress in this direction, following the lead given by Cohen and Levinthal (1989). (In this regard, see also Arora and Gambardella, 1994.)

9. This was a delicate issue as firms did not always understand the terms used. Quite unexpectedly, the term "*veille technologique*" (technology monitoring) raised particular problems! It may come as a surprise to find no question on the shortening of product-design times, which is one of the central elements of Mowery's analysis of R&D co-operation (1992). There are two reasons for this: the section on innovation objectives already included a question on the shortening of design times. Cross-classifying this question with R&D links should make it possible to confirm D. Mowery's conjecture or not: in fact, this objective seldom emerged in the preliminary interviews for the survey. Similarly, the advisability of risk-spreading was not mentioned as an objective. The model underlying this question therefore does not appear to be shared by firms.

10. In an ordered probit model, export ratios and the fact that R&D links are maintained with suppliers of materials and components significantly and positively influenced the share of turnover generated by innovation.

11. $$H = \frac{100}{\sum_{i}^{n} Sales_i} \times \sqrt{\sum_{i}^{n} (Sales_i)^2}$$

REFERENCES

ARCHIBUGI, D., S. CESARATTO and G. SIRILLI (1991), "Sources of Innovative Activities and Industrial Organization in Italy", *Research Policy*, Vol. 20, pp. 299-313.

ARORA, A. and A. GAMBARDELLA (1990), "Complementarity and External Linkages: The Strategies of the Large Firms in Biotechnology", *Journal of Industrial Economics*, Vol. 38, No. 4, June, pp. 361-379.

ARRORA, A. and A. GAMBARDELLA (1994), "Evaluating Technological Information and Utilizing it: Scientific Knowledge, Technological Capability, and External Linkages in Biotechnology", *Journal of Economic Behaviour and Organization*, Vol. 24, pp. 91-114.

BROUWER, E. and A. KLEINKNECHT (1994), "Determinants of Innovation and Diffusion: A Microeconometric Analysis in the Netherlands", paper presented at GENIST, Innovation, Patents and Technological Strategies, OECD, Paris, December.

CESARATTO, S and G. SIRILLI (1992) "Some Results of the Italian Survey on Technological Innovation", *STI Review,* No. 11, OECD, Paris, December, pp. 79-96

COHEN, W.M. and D.A. LEVINTHAL (1989), "Innovation and Learning: The Two Faces of R&D", *The Economic Journal*, Vol. 99, pp. 569-596.

De BRESSON, C. and F. AMESSE (1991), "Networks of Innovators: A Review and Introduction to the Issue", *Research Policy*, Vol. 20, pp. 363-379.

FARREL, J. and N. GALLINI (1988), "Second Sourcing as Commitment: Monopoly Incentives to Attract Competition", *Quarterly Journal of Economics*, Vol. 103, pp. 673-694.

FRIEDMAN, Ph., S. BERG and J. DUNCAN (1979), "External versus Internal Knowledge Acquisition: Venture Activity and R&D Intensity", *Journal of Economics and Business*, Winter, Vol. 31, No. 2, pp. 103-110.

HAGEDOORN, J. and J. SCHAKENRAAD (1991), "The Economic Effects of Strategic Partnerships and Technology Co-operation", Monitor/Spear, EC, Brussels, Luxembourg, September.

MOWERY, D. (1992), "International Collaborative Ventures and the Commercialization of New Technologies", in N. Rosenberg *et al.* (eds), *Technology and the Wealth of Nations*, Stanford University Press, Palo Alto, CA.

NAPOLITANO, G., "Industrial Research and Sources of Innovation: A Cross-Industry Analysis of Italian Manufacturing Firms", *Research Policy*, Vol. 20, pp. 171-178.

NELSON, R.R. and E.N. WOLFF (1995), "Factors behind Cross-Industry Differences in Technical Progress", Working Paper, Columbia University, May.

OECD (1992), "OECD Proposed Guidelines for Collecting and Interpreting Technological Innovation Data: Oslo Manual", OCDE/GD(92)26, Paris.

OECD (1994), *The Measurement of Scientific and Technological Activities: Frascati Manual 1993*, Paris.

PAVITT, K. (1984), "Sectoral Patterns of Technical Change: Toward a Taxonomy and a Theory", *Research Policy*, Vol. 13, pp. 343-373.

RIEDLE, K. (1989), "Demand for R&D Activities and the Trade-off Between In-house and External Research: A Viewpoint from Industry with Reference to Large Companies", *Technovation*, No 9, pp. 213-225.

SMITH, K and T. VILVEI (1992), "Innovation Activity and Innovation Outputs in Norwegian Industry", *STI Review*, No. 11, OECD, Paris, December.

TEECE, D. (1988), "Technical Change and the Nature of the Firm", in G. Dosi *et al.* (eds.), *Technical Change and Economic Theory*, Pinter Publishers, London.

HOW CAN VERY SMALL TECHNOLOGY FIRMS BE HELPED?

by
Philippe Kaminski
INSEE, Paris

INTRODUCTION

Over the past few years, firms in many countries have been asked to reply to a host of new surveys on innovation and related topics such as the transfer of technology or the protection of industrial property. We now have a clearer idea of the limits of such surveys, and some of the high hopes which had rather naively been voiced when the surveys were first launched have subsequently had to be scaled down substantially. Nonetheless, the information acquired from the surveys has proved to be highly instructive and we must endeavour to make the most of the lessons learned.

While the publication of statistical tables or academic studies is a necessary part of this learning process, it is not in itself sufficient. These surveys are expensive exercises and taxpayers are fully entitled to expect to see some benefit from them in economic terms; secondly, to encourage firms to reply to questionnaires, it was explained to them that the authorities needed the information in order to develop clearer policies for supporting innovation and technology diffusion. To show due respect for the time which firms were willing to spend on filling out questionnaires, this promise needs to be kept.

The work of the statistician is only properly completed when the figures he has produced help to inform and to provide insight into both the decisions made by government and the behaviour of major economic agents. We have therefore deliberately opted to stress practical rather than theoretical considerations in our contribution to this very timely discussion of the technological challenges facing very small firms.

UNDERSTANDING THAT A VERY SMALL FIRM IS NEITHER A BUDDING LARGE ONE NOR A SCALE MODEL OF A LARGE FIRM

Just as most big cities started out as tiny villages, most of today's large corporations, and especially those that have built their reputations on technological excellence and innovation, began as small workshops. There are exceptions, however, and more of them than one might think: some cities, like some businesses, were designed *ex nihilo* to become capitals or vast production and research centres from the outset. But society's imagination is captivated by stories of little guys who go from success to success until they turn into giants – so captivated that people come to believe that every small company is a budding large one, if only it can be given a boost. It is for this reason that aid or incentive programmes for small or very small businesses are almost always based, whether consciously or not, on a single objective: enabling them to expand or to join forces, or in any event to transcend their small size, which is implicitly viewed as a temporary condition or a handicap that needs to be overcome.

And yet, only a minority of small businesses ever expand. Moreover, not only do many of them not want to expand, but it could also be said that a fair number of them ought not to: they carry out their role within the diversity of the productive system and do it efficiently at the level that is theirs; they perform special functions that they could not discharge as well if they were larger – just like the tiny birds whose job it is to clean the crocodile's teeth. They will never grow big like the crocodile, and the crocodile needs them, just as they are, in order to survive and prosper.

A TOP PRIORITY: HAVING TECHNOLOGICALLY DEMANDING CUSTOMERS

A very small firm considers its greatest strengths to be flexibility and adaptability. It knows how to react quickly and how to deliver an appropriate response to even the most unexpected demands from its customers.

The overriding importance of the market – or, more exactly, of what customers want – emerges clearly from indicators of the stimulus to innovate. The sequence of events is obvious: the customer places an order, the firm manufactures the good, and lastly, in cases where the customer's request diverges from the norm, an innovation is produced.

The pace-setting (or even "bandwagon") effect triggered by competition is less significant than in-house interest in technology *per se*. The more innovative the sector, the wider the gap – the only exception to this being textiles, which is the least innovative sector but also the only one in which competitive pressures play a meaningful role in the diffusion of technology. In all cases, however, and even for firms specialised in process innovation, customer demand far outstrips other factors as the main engine for innovation.

To respond to these demands, micro-firms rely essentially on informal technological know-how. As a rule, the technical knowledge harnessed by craftsmen and their helpers has not been acquired formally, but gradually and in a manner not

Table 1. **Indication of the stimulus to innovate for different firm populations**[1]

	Customer demands	Internal technological dynamic	Competitive pressures
All firms	2.77	2.27	2.09
Production goods	3.11	2.63	2.26
Textiles and clothing	2.57	1.69	2.26
Significant product innovators	3.41	3.10	2.55
Product improvers	3.22	2.80	2.48
Non-innovators in products	2.49	1.94	1.88
Significant process innovators	3.26	2.88	2.35
Technology adaptors	3.23	2.74	2.37
Non-innovators in processes	2.45	1.94	1.85

1. The method used to obtain the findings presented below is described at the end of this chapter.
Source: Author.

clearly distinguishable from their productive activity – which is the defining feature of a genuine "trade". These technical skills, this tradesmanship, are the craftsman's only real capital – assets that are intimately connected with his individual identity; from this standpoint, any idea of expansion or even of working closely with another firm is regarded as a dilution of that capital, and a loss of freedom and identity.

It is remarkable to note how independent such attitudes are from the circumstances of the firms in which they are observed: neither age, activity, region, level of training nor ability to innovate has any effect on the values through which, as a group, craftsmen who carry out production define themselves.

The overriding importance of the "personal touch" and of passing down one's trade by word of mouth explains how hard it is to share or reproduce this capital in order to expand or to "network" micro-firms. On the contrary, their natural development involves a process of hiving off: new entities, of comparable size and operating in a similar manner, are created when the market so allows, thanks to an internal process by which the micro-firm's pattern of operations is appropriated. After a period of apprenticeship, which varies, those employees of small craft firms most capable of doing so go off on their own, and they, in turn, subsequently train the next generation of enterprise creators.

THE VERY REAL PROBLEM OF FINDING SKILLED PERSONNEL

It is a well known fact, however, that the French educational system still neglects vocational training and apprenticeship programmes. Inevitably, this causes micro-firms in manufacturing and technology enormous problems for finding dynamic staff who will be capable, in turn, of creating their own enterprises.

Table 2. **Is it easy for the firm to find skilled personnel who satisfy its technological needs?**[1]

	Yes		Yes, but only temporarily		No	
	%	Innovative potential	%	Innovative potential	%	Innovative potential
All firms	25.9	7.5	6.5	4.7	67.6	7.5
Minimum (textiles)	37.5	4.4	10.9	4.7	51.7	7.5
Maximum (climate engineering)	8.8	1.4	6.4	4.1	84.8	5.3
Production goods	29.2	10.4	6.5	5.6	64.3	10.5

1. The method used to obtain the findings presented below is described at the end of this paper.
Source: Author.

Overall dissatisfaction has hit an alarming level, since two out of every three firms consider the difficulty in finding good staff an impediment to their development, and to their technological development in particular. The problem is least acute in textiles (one firm in two), but that is the industry in which technology is least at issue. The extreme example of climate engineering can be seen as a cry for help.

It is significant that the minority of firms that checked the middle response – *i.e.* those patronising temporary employment agencies – were, in their respective sectors, among the least technologically involved, *i.e.* the least demanding in terms of know-how. This confirms the close link between technological capital, quality of staff and the staff's capacity to create new enterprises.

At this point, we already know what more appropriate government policies, and different behaviour on the part of economic agents, ought to bring micro-firms: customers – in other words, access to high-tech markets and an appropriate, efficient educational system that does not systematically divert the brightest people to theoretical studies or careers in big business. On this first point, special vigilance is necessary, as widespread pressures to scale back major programmes and defence spending bring about a tendency to steer an increasing share of government procurement towards large or very large firms.

Now we must describe what micro-firms do not need, and what it is therefore useless to try so hard to offer them.

ACCESS TO INFORMATION

Analysing the obstacles to innovation shows that, in most cases, what holds micro-firms back is their desire to stick to their own trade, to the technological skills in which they excel, and not to embark upon a course of risk-ridden diversification.

Table 3. **Breakdown of firms by leading impediment to innovation**[1]

	Desire not to have to expand	Desire to stick to the same trade	Not knowing where to find information	No response
All firms	24.7	41.1	12.6	15.6
Production goods	28.2	32.6	17.1	22.2
Textiles (minimum column 1)	10.3	44.4	11.8	33.5
Climate engineering (minimun column 3)	27.5	47.9	7.2	17.4

1. The method used to obtain the findings presented below is described at the end of this paper.
Source: Author.

It is fairly difficult to imagine, however, how government policies could get firms to agree more readily to change their trades. It would probably be preferable to respect their wishes and help them find innovative outlets in their own specialities, and to rely on the appearance of new firms to adapt the productive mechanism to broad shifts in demand trends.

The second obstacle cited was the desire not to expand; this argument, which we have already mentioned, should not be rebutted out of some preconceived notion of what others ought to think. On the other hand, the obstacle cited last unfortunately corresponds to services that government knows how to provide and that it is naturally trying to develop: aid in searching for the right information. And yet, only 17 per cent of firms in the Production goods sector explicitly feel a need for this, and the proportion drops to less than 13 per cent for all manufacturing or technology micro-firms as a whole.

Moreover, scientific and technical information (STI) is rarely cited as a source of innovation, especially if the information is organised, *i.e.* if it costs money (specialised workshops, data bank servers, CD-ROM, etc.):

Table 4. **Comparative intensity of innovation sources**[1]

In-house R&D	Proto-R&D	Outside R&D	Licences	Organised STI	Informal STI	New machinery	New staff	New materials
750	2 020	400	110	650	890	930	570	990

1. The method used to obtain the findings presented below is described at the end of this paper.
Source: Author.

The predominance of in-house know-how is overwhelming; but in fact, all of the four most frequently cited sources are highly unlikely to be promoted by any government aid or service scheme: "proto-R&D", that is to say all methodological or experimental activities integrated into the production process which have the same goal as formal R&D but which cannot be dissociated from the other functions of the firm; new materials (and components), which are more often imposed on the firm either by customers or by circumstances than adopted as a result of the acquisition of specific information; new machinery, which firms are obliged to learn how to use in order to meet productivity demands or because the older models can no longer be maintained; lastly informal scientific and technical knowledge, which is provided by customers, suppliers or the trade and which is an integral part of the day-to-day business dealings of the firm.

DIRECT SUPPORT FOR INNOVATION

On the whole, craftsmen feel that these government aid and service schemes, which they help to pay for with their taxes, are designed for large firms and ill-suited to their own needs; as a result, they consider that such programmes put them at a disadvantage. And yet some schemes are oriented in ways that favour small and medium-sized industries (SMIs); even so, an SMI has to have enough of a formal organisation to be able to spare a manager's time to prepare an application and defend it, and it has to be sufficiently diversified for its projects, as well as the corresponding pay-back periods, to be clearly defined.

Even when simplified, these constraints do not fit easily into the natural process by which innovation emerges in a micro-firm: customers place orders, craftsmen produce, and sometimes innovation is the result. As a rule, there is no prior desire to innovate, but a harnessing of know-how in new or unexpected situations which will subsequently, and necessarily, lead to more or less radical innovations.

In contrast, entering into partnership with a government aid agency presupposes the existence of an idea, or of something being developed, but without a clear-cut outlet (just hopes for a future market) such that carrying through with it would be a big enough gamble or enough of a technological feat as to require funding and advice.

While this is not absolutely incompatible with the operation of a micro-firm, it is plain to see that such situations are not commonplace. Analysing attitudes towards funding difficulties leads to similar conclusions: micro-firms tend not to give innovation a special place with respect to their overall financial problems – and for the same two reasons: first, because they consider innovation to be more of an outcome than an objective; and second, because that sort of bank financing would appear to be reserved for larger businesses.

Does this mean there can be no direct support for the innovative activities of micro-firms? Certainly not, as long as procedures are simple and quick and they reach their targets. Take patenting, for example: while just 10 per cent of micro-firms have already registered a patent, only a little more than 50 per cent say the subject is irrelevant to them.

Thus a quarter of all firms – and the more innovative the sector, the higher the proportion – consider that their access to patents is blocked because the system is too costly. This should suggest some simple and selective aid measures.

Table 5. **Attitudes towards patenting**[1]

	Too costly	Protection illusory	Irrelevant
All firms	26.0	21.9	52.1
Textiles	12.7	40.8	46.5
Electrical services (max. col. 3)	23.4	7.8	68.8
Production goods	36.7	27.5	35.8

1. The method used to obtain the findings presented below is described at the end of this paper.
Source: Author.

The same holds true for R&D, the definition of which needs some adjustment to fit in with the realities of the micro-firm: 12 per cent of such firms say they engage in it "regularly" and 18 per cent "occasionally". A flat-rate system modelled on research tax credits would probably prove an effective incentive at modest cost.

Table 6. **Average indicator of innovation performance, by R&D and patenting experience**[1]

	Have registered a patent	Interested in registering a patent	Not relevant	All firms
Regular R&D	19.0	14.6	14.3	15.3
Occasional R&D	12.4	10.9	8.1	9.6
No R&D	9.8	8.4	4.5	5.3
All firms	13.4	10.4	5.5	7.3

1. The method used to obtain the findings presented below is described at the end of this paper.
Source: Author.

The more a firm is involved in R&D, the greater the interest it takes in patents and the greater its capacity for innovation. Although the conepts involved are difficult to describe in terms of cause and effect than they are for large firms, they are nonetheless remarkably clear in that the correlation can be seen not only in the totals but also within each individual line and column. The few micro-firms in the sample that perform R&D on a regular basis and that have already filed a patent score, on average, 19 out of a possible maximum of 20 points, *i.e.* they all have the highest level of technological excellence according to all criteria. The average general score is 7.3, and it is clear that a median position in either of the two indices would be sufficient for a firm to achieve a higher score (*i.e.* higher than the average); a firm performing R&D occasionally but with no interest in patents would score 8.1, whereas a firm with a mild interest in patents but with no R&D activities would score 8.4.

ARE MICRO-FIRMS FUNDAMENTALLY DIFFERENT FROM SMIS?

We have highlighted disparities between SMIs and micro-firms, but also things they have in common. A comparison of the sources of innovation provides some interesting insight into both.

Table 7. **Sources of innovation (SMIs): results of the 1991 industrial survey by size category**[1]

	In-house R&D	Proto-R&D	Patents registered	Outside R&D I[2]	Outside R&D II[3]	Licences	New machinery	New materials
20 to 49	624	997	248	284	476	186	1 016	727
50 to 99	970	1 300	391	510	563	270	1 189	839
100 to 199	1 302	1 566	495	722	599	297	1 356	965
200 to 499	1 713	1 715	657	987	716	405	1 460	1 022

1. The method used to obtain the findings presented below is described at the end of this paper.
2. Outside R&D I: Undertaken by an affiliated firm.
3. Outside R&D II: Undertaken by an unaffiliated firm.

Source: Author.

The intensity of each source diminishes with the size of a firm, but the rate of decrease can vary widely from one source to another. The effect of new machinery tapers off very gradually, meaning that its relative importance increases as firms get smaller. The same holds true for in-house know-how (proto-R&D) and, to a lesser extent, new materials.

In contrast, the impact of R&D is very strongly correlated with the size of a firm. Above 500 employees, in-house R&D reigns supreme; at 500 employees, it drops to the same level as proto-R&D; at 200 employees, to that of new machinery; and at 50 employees, to below that of new materials. Unfortunately for SMIs, let alone micro-firms, R&D is what is aided best.

Nonetheless, while proto-R&D is a highly individual matter for a firm and cannot possibly be assessed from the outside, an imaginative industrial policy should devise aid systems tailored to the effects of each technological input.

Table 8. **Sources of innovation: an attempt to link up the two surveys**[1]

	In-house R&D	Proto-R&D	Outside R&D	Licences	New machinery	New materials	Total
10	750	2 020	400	110	930	990	5 200
20 to 49	624	997	380	186	1 016	727	3 930
50 to 99	970	1 300	537	270	1 189	839	5 105
100 to 199	1 302	1 566	660	297	1 356	965	6 146
200 to 499	1 713	1 715	851	405	1 460	1 022	7 166
500 to 999	2 300	2 043	1 208	490	1 772	1 318	9 131
1 000 to 1 999	2 702	2 353	1 326	585	1 827	1 439	10 232
2 000 or more	2 961	2 451	1 614	690	2 157	1 742	11 615

1. The method used to obtain the findings presented below is described at the end of this paper.
Source: Author.

It can be seen that these inputs have been reversed: whereas in SMIs the influence of materials and components is always less than that of equipment, in micro-firms it is slightly greater. Micro-firms often have a minimal amount of equipment – probably due to cost, but also because the priority they give to know-how is more compatible with older equipment. For this reason, the computer facilities of many micro-firms would be deemed inadequate from the outside, whereas the craftsmen using them do not seem in any way inconvenienced. Consequently, aid mechanisms will have to direct their creativity more towards new materials and components: it is their turn to innovate.

SOME PURELY DOMESTIC FRENCH CONSIDERATIONS

While craftsmen are frequently unswerving critics of the inefficiency of government aid and advisory agencies, they actually know little about them and, on average, are able to name just over two [apart from ANVAR (*Agence nationale de valorisation de la recherche*), which came under a different heading in the survey]. From this, it is obvious that firms have to be approached, and help made accessible to them.

But who should be responsible for doing this? We have estimated that in France there are just over 20 000 craft industries that play a role in the diffusion of technology and, in some cases, the production of innovation. In some ways they are comparable to SMIs; in others, they are markedly different.

In France, such businesses fall under Chambers of Trade and the Ministry of Crafts; the emphasis is therefore more on what sets them apart than on their similarities to industrial firms. Yet within the world of small businesses they are lost in a sea of 800 000 bakeries, hairdressing salons and plumbing businesses with which they have little in common.

This situation is not conducive to thorough review or a coherent reform of existing mechanisms. While we do not claim to offer a solution, we nonetheless feel that a focus on technological aspects, addressing the entire manufacturing sector yet capable of providing solutions tailored to the special characteristics of the smallest businesses, would be better than a scattered approach and, in any event, better able to overcome the inertia that hampers the effectiveness of industrial policies.

GENERAL PRESENTATION OF THE "TIME" SURVEY AND ITS MAIN FINDINGS

All of the findings presented in this article are taken from the results of a survey on technology and very small businesses: TIME. "TIME" is the French acronym for "Technology and Innovation in the Micro-Firm". The idea for it took shape shortly after publication of the first articles on the survey on industrial innovation carried out in France in 1991; since that survey was restricted to companies having more than 20 employees, it was only natural to wonder about the behaviour of smaller firms.

The TIME survey was conducted in two phases (autumn 1992 and spring 1993), as a partnership between INSEE (*Institut supérieur des métiers*) and EUROSTAT. In all, it involved 1 016 micro-firms in manufacturing or maintenance located in five French regions: Centre, Midi-Pyrénées, Franche-Comté, Nord and Ile-de-France. Data was collected by specially trained pollsters, the average interview lasting just over an hour. The sample was drawn from a parent population estimated at 21 000 firms nation-wide, defined by a group of three criteria: engaged in the same line of business for at least the past five years, listed in the trade directory, and turnover of more than FF 2 million (it was deemed preferable to base the size threshold on turnover, which is less subject to fluctuation than the number of salaried employees).

The questionnaire incorporated all headings in the Oslo Manual that were meaningful for micro-firms; these were supplemented by questions on the behaviour of craftsmen and their attitudes towards technology, as well as by questions of purely domestic French interest on their institutional and financial environment.

Below a threshold set theoretically at ten employees, but which in fact can vary, manufacturing firms in France fall under the vast trades sector, which has its own institutions and elected representatives. We wanted to avoid certain words that were too exclusively associated with a French context, preferring the more recent term "micro-firm" in order to facilitate international comparisons. On the other hand, because of space constraints we used the word "craftsman" (*artisan*) instead of "head of micro-firm".

Summary findings of the TIME survey were published in September 1994 in No. 41 of the SESSI (*Service d'études des stratégies et des statistiques industrielles*) publication; more detailed findings were compiled in a folder presented at a colloquium on 4 October 1994 organised by the University of Paris XII-Créteil.

METHODOLOGICAL NOTES ON THE TABLES OF FINDINGS

The findings presented in this article are either proportions or averages, computed from the population of respondents (1 016 units) using a weighting system to ensure that the outcome was representative of France as a whole. The weighting was intended to correct variations in response rates for the regions in which the survey was conducted, inter-regional variations in the structure of activities, as well as the quality of interviews (some businesses, wishing to cut the discussion short, claimed that they engaged in no innovation, which could not always be clarified by telephone at the moment data were entered). Yet the weightings were rarely far from 1, and it may be considered, for purposes of initial interpretations, that the data are raw.

There are two different categories of averages.

1. One type involves questions presented in the questionnaire in the form of rankings. Such is the case for the sources of and stimulants to innovation (Tables 1, 4, 7 and 8), which are scaled logarithmically, with the average result theoretically able to range between 0 and 4:

- no influence: 0;
- little influence: 1;
- moderate influence: 2;
- strong influence: 4.

For tables relating to sources (4, 7 and 8), the outcome was multiplied by 1 000 in order to retain four digits.

2. The other type involves an index that aggregates several different questions (Tables 2 and 6). Calculation formulae are based on the ones developed in connection with the Industry survey: a number of indicators are aggregated, in order to allow strong areas to offset weak ones, so that a single scale of ranking emerges. For each firm, the range of possible outcomes goes from 0 to 20 – which is not to say that a score of 10 is in any way central or discriminant; that depends entirely on the conventions of the calculation.

The indicator used in Tables 2 and 6 summarises all items directly connected with innovation: product/process, precursor/imitator, major/incremental, completed/under way/in the preparatory phase. The theoretical maximum score of 20 is given to firms that come out on top on all counts. Processing the TIME survey has given rise to the calculation of other indicators of this sort, relating to computerisation, R&D, interest in scientific and technical information and receptiveness to outside partnerships.

(continued on next page)

(continued)

Special note on Table 1

We had to reconsider the traditional opposition between market pull and technology push, since the size and economic impact of micro-firms make it difficult for them to have an overall, unifying perception of the market, *i.e.* of the forces of supply and demand. Micro-firms are, however, perfectly capable of identifying their customers' needs, as well as what their direct competitors are able to deliver. This is why we have split market pull into two distinct categories; and the responses show clearly that they are indeed separate, since one was ranked higher than technology push, the other lower.

For this table we included only two sectors of activity, those located at opposite ends of each of the indicators of innovation: all manufacturers of production goods, a sector ranked as the most highly innovative regardless of the indicator selected; and textiles and clothing, a sector in which the constant quest for novelty and the presence of rather heavily capital-intensive firms do not prevent it from being the least innovative sector, in the technological sense of the term.

Special note on Table 8

Linking together two different surveys is always dangerous, even if, as in this case, questions were coded identically. The survey on industry was conducted through the mail, the one on micro-firms through visits; in the first case, totally non-innovative firms all had the same score of zero for intensity of sources, whereas in the second survey they were able to respond if they so desired. This leads to high discontinuity in the "Total" column and in the "Proto-R&D" column, the predominance of proto-R&D being inflated by the statements of the least technologically active firms.

EMBODIED AND DISEMBODIED INNOVATIVE ACTIVITIES: EVIDENCE FROM THE ITALIAN INNOVATION SURVEY[1]

by

Rinaldo Evangelista

Institute for Studies on Scientific Research, National Research Council, Rome

Science Policy Research Unit, University of Sussex, Brighton

INTRODUCTION

Schumpeter (1934, 1939, 1942) rediscovered and emphasized the importance of technological change for the dynamics of growth and competition. While the classical economists (Smith, Ricardo and Marx) regarded technological change as largely embodied and related to accumulation of capital and investment activities of firms, more recent contributions, especially of an empirical nature, have tended (explicitly or implicitly) to view technological change as largely disembodied and related to the production of new knowledge and innovation.

Investment in new machinery has thus progressively lost its central position in the analysis of technological change, and increasing attention has been given to innovative activities aimed at producing new technological knowledge, as the technological indicators used in empirical studies clearly show. The innovative performance and technological profiles of countries, sectors and firms are usually measured using "disembodied technology indicators", such as R&D, patents, licences, and innovation counts, which measure innovative activities aimed at producing new ideas. Instead, indicators of innovative efforts to use and diffuse these new technologies through investment in new process technologies have not been developed and are not been generally emphasized.

Recently, however, investment seems to have regained a place in the research agenda (Scott, 1988). It has been found to play an important role, along with other technological inputs, in explaining cross-country differences in growth rates (Dosi *et al.*, 1990; Maddison, 1991; Amendola *et al.*, 1993; Pianta, forthcoming). Basic differences among industrial sectors and firms have also been described in terms of sources of technology, among which efforts to improve process technologies through R&D and investment have been identified as important, albeit in different degrees for different sectors (Pavitt, 1984). However, how much investment, as a distinct source of technology, matters in terms of a firm's innovative behaviour and what its relationship is to disembodied innovative efforts have not been fully investigated. These issues are explored here, using data provided by the CNR-ISTAT (National Research Council of Italy-Italian National Statistical Institute) innovation survey. The latter has provided unique information on the innovative activities of a very large sample of Italian firms. These data allow us to:

- show the importance in firms' innovative behaviour of investment in new machinery and equipment as compared to the more disembodied forms of innovative activities represented by R&D and design and engineering;
- explore whether the relationship between disembodied and embodied technological activities at firm and sectoral level are complementary or substitutive;
- identify main sectoral patterns of innovation in terms of disembodied/ embodied innovative activity, with respect to firm size and labour productivity.

DESCRIPTION OF THE DATABASE AND VARIABLES USED IN THE STUDY

The empirical work reported in this article refers to Italy and uses data provided by the CNR-ISTAT survey of innovation in Italian manufacturing industry for 1981-85.[2] The database provides information on total costs of innovation for 8 220 innovating Italian firms undertaking a wide range of technological activities. Total innovation costs include expenditures on R&D, design and engineering (D&E), innovative investment (INV), and innovative marketing activities. These data have been matched with 1985 data from the survey of gross industrial product survey in order to have, for a slightly smaller sample of firms (6 839), data on sales, total investment in machinery, and value added.[3]

The disaggregation of total innovation costs provided by the CNR-ISTAT database allows us to distinguish between innovative activities devoted to the "production" (and development) of new knowledge, as measured by R&D and D&E activities, and those devoted to the "use" and "application" of such technology on an industrial scale, as measured by the innovative investments. We label the former "disembodied" and the latter "embodied".[4] Quantitative measures of disembodied innovative intensity are built by dividing R&D and D&E expenditures by number of employees. Embodied innovative intensity will be represented by innovative investment per employee.

Indicators of the internal composition of innovative activities are represented by shares of R&D, D&E (for disembodied innovation), and innovative investment (for embodied innovation) in total innovation costs.

Total investment in machinery per employee (INVMAC) is used as a proxy for the capital intensity of the production process and value added per employee (VA) as a proxy for labour productivity. Firm size will be measured by sales (SALES). Table 1 shows the variables used in the statistical analysis of the following sections.

Table 1. **List of variables**

Industrial structural variables	SALES INVMAC VA	Sales per firm (billion lire) Investment in machinery per employee Value added per employee
Innovative intensity variables	INCOST	Total innovation costs per employee
disembodied	R&D D&E RDDE	R&D expenditure per employee Design and engineering (D&E) expenditure per employee R&D + D&E
embodied	INV	Innovative investment expenditure per employee
	MKT	Innovative expenditure per employee in marketing
Innovative composition variables[2]	% R&D % D&E	Per cent of R&D innovation cost on total innovation costs Per cent of D&E innovation costs on total innovation costs
	% INV	Per cent of innovative investment on total innovation costs
	% MKT	Per cent of innovative expenditure in marketing on total innovation costs

1. INCOST = D&E + INV + MKT.
2. % R&D + % D&E + % INV + % MKT = 100.

Source: Author.

THE RELATIVE IMPORTANCE OF EMBODIED AND DISEMBODIED INNOVATIVE ACTIVITIES

Table 2 shows the average values of the economic and innovation variables listed in Table 1 for total manufacturing industry and for 30 industrial sectors. The average values for the composition of innovative activities and innovative intensity show the importance of innovative investment expenditure. The latter represents 53 per cent of total innovation costs and approximately one-third of the total investment in machinery. Also interesting is the relatively small share of R&D activities in total innovation costs. On average, R&D expenditures are as little as 17.4 per cent of firms' total innovative efforts. Also, the percentage devoted to D&E activities is quite low when compared to the embodied component of innovative activities. Innovative efforts related to marketing have the smallest share.

Table 2 highlights the large differences among industries in the variables related to the composition and intensity of innovative activities. This indicates, in fact, that a wide variety of sectoral innovative patterns appear when the various components of innovative activities are taken into account.

Table 2. **Innovation and industrial structure variables by industrial sector**

Average values

Industrial sectors	Firm size	Capital intensity[1]	Labour productivity[1]	Innovative intensity[2]					Innovative composition			
	SALES	INVMAC	VA	INCOST	R&D	D&E	INV	MKT	% R&D	% D&E	% INV	% MKT
Energy and gas	1 469	25.9	107.2	3.66	0.52	0.46	2.63	0.06	14.2	12.5	71.8	1.5
Metals	124	9.8	40.3	1.79	0.17	0.30	1.22	0.10	9.6	16.9	67.9	5.6
Non-metals, minerals	16	6.2	43.7	3.90	0.19	0.34	3.26	0.11	4.9	8.7	83.7	2.7
Chemicals	89	8.9	56.3	2.72	0.64	0.42	1.50	0.17	23.3	15.6	55.0	6.1
Pharmaceutical	65	5.9	66.7	3.57	1.40	0.58	1.15	0.44	39.3	16.2	32.2	12.2
Synthetic fibres	222	11.0	49.0	2.22	0.26	0.22	1.69	0.05	11.6	10.1	76.1	2.1
Metal products	11	4.2	40.6	2.08	0.14	0.34	1.53	0.07	6.6	16.3	73.5	3.6
Gen. mechan. engin.	22	3.5	46.3	2.50	0.41	0.76	1.22	0.11	16.5	30.4	48.7	4.4
Metal machinery	10	3.3	43.1	3.51	0.72	1.36	1.28	0.14	20.6	38.7	36.6	4.1
Textile machinery	16	3.8	39.8	4.74	0.37	2.15	2.06	0.17	7.7	45.2	43.5	3.5
Office mach., comp.	786	17.1	107.2	14.40	4.99	3.85	3.56	2.00	34.7	26.8	24.7	13.9
Electrical comp.	26	3.3	43.5	1.89	0.30	0.47	1.05	0.07	16.0	24.7	55.7	3.6
Electronic eq. and comp.	62	4.2	48.3	6.56	2.13	2.72	1.47	0.24	32.5	41.5	22.4	3.6
Radio, TV and comm.	56	8.0	46.2	7.64	1.65	2.35	3.34	0.29	21.7	30.8	43.7	3.8
Electrical appl.	32	2.8	27.4	1.56	0.17	0.34	0.91	0.14	10.7	21.7	58.4	9.1
Motor vehicles comp.	25	4.5	37.6	2.80	0.43	0.68	1.62	0.06	15.5	24.3	58.0	2.2
Motor vehicles	1 936	6.4	44.3	6.38	0.98	1.50	3.79	0.12	15.3	23.5	59.4	1.8
Other transport	43	2.0	28.1	1.18	0.18	0.33	0.62	0.05	15.2	27.9	52.3	4.6
Aircraft	174	4.6	47.1	13.32	0.31	6.68	6.32	0.01	2.3	50.2	47.4	0.0
Precision instruments	12	4.2	41.2	5.13	1.06	1.39	2.27	0.41	20.7	27.1	44.2	8.0
Food	42	7.3	44.3	2.77	0.10	0.17	2.37	0.13	3.6	6.2	85.5	4.6
Sugar, drinks	96	6.2	48.9	1.83	0.20	0.24	1.33	0.06	11.0	13.3	72.6	3.2
Textiles	15	4.7	40.5	2.05	0.13	0.15	1.71	0.06	6.4	7.1	83.5	3.0
Leather	11	3.0	38.8	1.55	0.06	0.15	1.28	0.05	3.8	10.0	82.7	3.3
Footwear, clothing	15	1.6	33.8	0.69	0.04	0.13	0.49	0.05	5.4	18.0	70.0	6.6
Wood, furniture	7	3.3	36.1	2.02	0.16	0.24	1.50	0.12	7.7	12.1	74.4	5.8
Paper, printing	28	5.8	55.0	2.91	0.13	0.13	2.52	0.13	4.4	4.3	86.6	4.6
Plastic	13	5.8	46.9	3.29	0.31	0.56	2.25	0.16	9.5	17.0	68.6	4.9
Rubber	34	5.0	37.3	8.46	3.17	2.42	1.69	1.18	37.5	28.7	19.9	13.9
Other manufacturing	10	3.9	36.0	2.04	0.26	0.32	1.38	0.07	12.9	15.6	67.9	3.6
Inter-sect. coef. of var.	2.43	0.78	0.38	0.83	1.49	1.34	0.60	1.75	0.70	0.55	0.33	0.68
Total	35	5.9	46.8	3.62	0.63	0.87	1.93	0.19	17.4	24.1	53.3	5.2

1. In millions of Italian lire in 1985.
2. In millions of Italian lire per employee.

Source: Based on the CNR-ISTAT database.

Innovative investment (INV) emerges as a very important source of technology in all sectors. Innovative investment per employee is high both in certain R&D-intensive sectors (*e.g.* office machinery, computing, precision instruments, radio/TV and communication components) and also in sectors with low R&D and D&E innovative intensity, such as non-metals, food, wood and furniture, textiles, paper and printing, and plastics. The large shares of innovative investment (% INV) for most of the traditional consumer goods sectors show that they rely heavily upon external technological sources even if the effective expenditure per employee is below average. On the other hand, metals and machinery show low R&D intensity and a much higher D&E intensity and composition. D&E intensity is also high (over 2 million lire per employee) in sectors such as textile machinery, office machinery, computing, radio-TV and communication components, aircraft-aerospace, and rubber. Industrial sectors with high R&D intensity (from 1 to 5 million lire per employee) are pharmaceuticals, office machinery, computing, electronic equipment and components, precision instruments, and rubber.

Table 2 also shows wide sectoral differences in firm size and, to a lesser extent, in capital intensity and labour productivity. Traditional industries (*e.g.* textiles, leather, footwear and clothing, wood and furniture) and specialised sectors (*e.g.* specialised machinery, general mechanical machinery, metal machinery, textile machinery, and precision instruments) are characterised by small and medium-sized firms and low capital intensity. Sectors characterised by continuous production processes and for which scale factors are relevant, such as energy, metals, synthetic fibres, chemicals, rubber and plastics, and motor vehicles, are characterised instead by medium-high capital intensity and large firms. Among the highly innovative sectors, office machinery shows a surprisingly high level of investment in machinery per employee.

Along with sectors, firm size also represents an important determinant of innovative patterns. Table 3 presents the data broken down by firm size and sales. All indicators of innovative intensity (both disembodied and embodied) increase with firm size, with small and large firms showing greater differences in R&D and D&E (Archibugi *et al.*, forthcoming). Differences in the composition of innovative activities also appear among firms classed by size. Spending on R&D accounts for 20 per cent for the largest firms and 12 per cent for the smallest; this confirms that this indicator underestimates the innovative efforts of the smaller firms. The pattern is the same for the D&E indicator. Conversely, the relative weight of innovative investment is inversely related to firm size; this confirms the higher propensity of this category of smaller firms to concentrate most of their innovative efforts in technologies embodied in machinery and equipment. These differences are thus related to the sectoral evidence presented in Table 2.

Both Tables 2 and 3 point to the danger of relying exclusively on R&D expenditure as the key indicator of innovative activities in manufacturing industry and confirm the need for a broader view of innovative efforts that includes D&E and, especially, investment. More particularly, they show that for economies where there

Table 3. **Innovation and industrial structure variables by firm size**

Average values

Firm size (billion lire)	Capital intensity[1] INVMAC	Labour productivity[1] VA	Innovative intensity[2]					Innovative composition			
			INCOST	R&D	D&E	INV	MKT	% R&D	% D&E	% INV	% MKT
Up to 100	4.6	42.8	2.74	0.33	0.48	1.80	0.13	12.2	17.4	65.6	4.8
100 to 500	5.9	50.2	3.45	0.68	0.83	1.81	0.13	19.9	24.0	52.5	3.7
Over 500	7.8	50.3	5.06	1.04	1.50	2.21	0.31	20.6	29.6	43.6	6.2
Total	5.9	46.8	3.62	0.63	0.87	1.93	0.19	17.4	24.1	53.3	5.2

1. In millions of Italian lire in 1985.
2. In millions of Italian lire per employee.

Source: Based on the CNR-ISTAT database.

are more smaller than larger firms, relying on R&D expenditure as an indicator of innovative activity seriously underestimates innovative activity and overestimates the importance of R&D as a source of innovation.

THE RELATIONSHIP BETWEEN EMBODIED AND DISEMBODIED TECHNICAL CHANGE AT FIRM LEVEL

We use the data set to investigate complementary or substitutive relationships between disembodied (R&D and D&E) and embodied (INV) innovative activities at the firm level. Table 4 shows the correlation coefficients between the indicator of disembodied innovative intensity (sum of R&D and D&E – RDDE in the table), innovative investment (INV), firm size (SALES), capital intensity (INVMAC) and labour productivity (VA). All correlation coefficients have been computed using logarithmic values.[5]

Table 4. **Inter-firm correlation (matrix log values)**[1, 2]

	LSALES	LINVMAC	LVA	LINCOST	LRDDE	LINV
LSALES	1.000					
	P = .					
LINVMAC	0.309	1.000				
	P = .000	P = .				
LVA	0.368	0.387	1.000			
	P = .000	P = .000	P = .			
LINCOST	0.050	0.342	0.270	1.000		
	P = .001	P = .000	P = .000	P = .		
LRDDE	0.082	0.203	0.248	0.803	1.000	
	P = .000	P = .000	P = .000	P = .000	P = .	
LINV	0.006	0.358	0.215	0.891	0.514	1.000
	P = .348	P = .000	P = .000	P = .000	P = .000	P = .

1. The first row shows the correlation coefficients and the second row the statistical significance.
2. Number of cases considered: 3 734.

Source: Based on the CNR-ISTAT database.

Analysis of the correlation matrix shows clear complementarity between embodied and disembodied innovative activities. R&D and D&E expenditures per employee (RDDE) are positively correlated to innovative investment.[6] The link is much weaker, even if still very significant, when we use total investment in machinery (INVMAC) as a proxy for embodied innovative activities.

Table 5. **Inter-firm factor analysis**

Variable	Communality	Factor	Eigenvalue	% of variation	Cummulative %
Principal components extracted: 2 factors					
LSALES	.69900	1	2.09813	42.0	42.0
LINVMAC	.55365	2	1.20949	24.2	66.2
LVA	.60465				
LRDDE	.67903				
LINV	.77128				

	Factor 1	Factor 2
Rotated factor matrix		
1. LSALES	.82130	–.15640
2. LINVMAC	.65156	.35933
3. LVA	.74479	.22349
4. LRDDE	.10600	.81719
5. LINV	.09750	.87280

Source: Author.

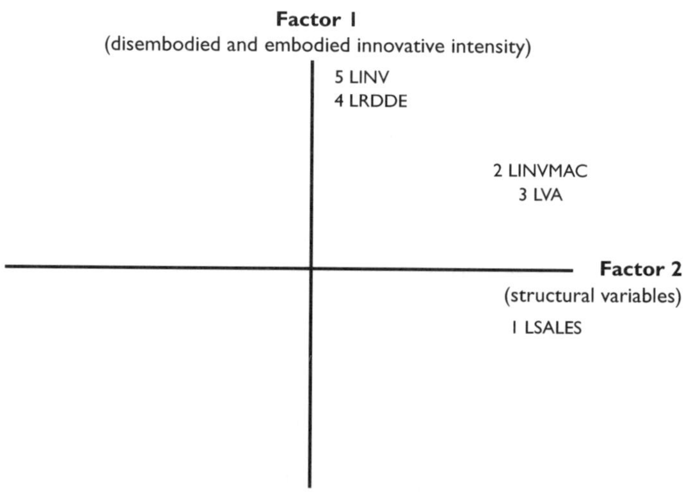

Source: Author.

These two results, together with the relatively low correlation between total investment in machinery per employee (INVMAC) and innovative investment (INV), have the following important implication: total investment in machinery is an imperfect proxy for a firm's embodied innovative efforts; in other words, the technological content of total investment in machinery varies greatly among firms. Total investment in machinery is more related to the organisational and structural aspects of production activities. These include the capital-intensive nature of production processes, to which large-scale production and higher levels of labour productivity could be associated.

Both the complementarity between embodied and disembodied technological activities and the structural dimension of production processes, represented by the three economic variables, namely, firm size (SALES), capital intensity (INVMAC) and labour productivity (VA), are effectively summarised by a factor analysis carried out using the same variables (Table 5). The first two factors explain 66 per cent of the total variance. Table 5 shows the rotated correlation matrix and the position of the different variables with respect to the two factors (after rotation). The first factor represents embodied and disembodied innovative intensity variables, while the second summarises production-related linkages among firm size, capital intensity, and labour productivity.

The relatively low values of the correlation coefficients in Table 4 and in the factor analysis reflect high inter-firm variance. This is due to two factors. The first is the large size of the sample, which makes all coefficients statistically significant even when correlation coefficients are rather low. The second is the non-selective nature of the sample, which covers all industrial sectors and different size classes, which, as already shown, are characterised by different technological profiles.

SECTORAL INNOVATIVE PATTERNS BASED ON EMBODIED AND DISEMBODIED TECHNICAL CHANGE

In this section, we analyse major sectoral innovative patterns on the basis of the disembodied-embodied nature of innovative activities. The results of regressions with sectoral dummy variables (LSDV) indicate the existence of fixed sectoral effects. Even using dummy variables for a rather small number of sectors, the results reported in Table 6 show that sectoral effects on the intensity of disembodied innovative activities are generally very significant, reflecting marked differences among sectors in the levels of technological opportunities (for generating new knowledge and innovations), and are independent of embodied innovative efforts. On the other hand, fixed sectoral effects are less important for explaining the level of value added per employee when both innovative and structural variables are taken into account. Value added per employee is significantly linked to firms' (embodied and disembodied) innovative efforts, capital intensity and size,

Table 6. **Multiple regressions with sectoral dummy variables**

Independent variables	Dependent variables: LRDDE T	LRDDE Sig. T	LVA T	LVA Sig. T
Constant	–6.34	0.0000	47.01	0.0000
Sect. dummy variables				
Energy and gas	–3.96	0.0001	5.77	0.0000
Metals	–6.70	0.0000	1.88	0.0597
Non-metals, minerals	11.55	0.0000	1.45	0.1466
Chemicals	0.73	0.4632	10.24	0.0000
Synthetic fibres	–0.99	0.3208	0.92	0.3588
Metal products	–8.24	0.0000	–0.24	0.8083
Mechan. engin.	0.56	0.5738	3.69	0.0002
Office mach., comp.	4.06	0.0001	0.68	0.4940
Motor vehicles	–2.67	0.0077	–3.62	0.0003
Precision instruments	1.38	0.1680	–0.23	0.8184
Food	10.24	0.0000	–0.10	0.9219
Sugar, drinks	–5.69	0.0000	3.13	0.0018
Textiles	10.10	0.0000	0.06	0.9560
Leather	–4.44	0.0000	0.18	0.8575
Footwear, clothing	–8.56	0.0000	–4.82	0.0000
Wood, furniture	–8.60	0.0000	–3.17	0.0015
Paper, printing	–9.36	0.0000	3.32	0.0009
Rubber, plastics	–7.41	0.0000	2.09	0.0370
Other manufacturing	–4.41	0.0000	–1.98	0.0475
LINV	40.13	0.0000	3.24	0.0012
LINVMAC			14.73	0.0000
LSALES			14.84	0.0000
LRDDE			7.05	0.0000
R^2		0.367		0.302
F		107.6		69.8
P.		0.0000		0.0000

Source: Based on the CNR-ISTAT database.

suggesting that all these factors contribute to explaining sectoral differences in labour productivity.[7]

Without analysing the correlation matrix, we can describe briefly intersectoral links among the different dimensions of innovative activities by applying the results of the factor analysis to a larger number of innovation variables (innovative composition variables have also been included), using average values for 104 industrial sectors.[8]

Table 7. **Inter-sector factor analysis (three-digit – 104 sectors)**

Variable	Communality	Factor	Eigenvalue	% of variation	Cummulative %
Final statistics					
SALES	.36319	1	3.01522	37.7	37.7
INVMAC	.81718	2	2.41835	30.2	67.9
VA	.65861				
RDDE	.75154				
INV	.49532				
YR&D	.66576				
YD&E	.70966				
YINV	.97231				

	Factor 1	Factor 2
Rotated factor matrix		
8. % INV	–.98039	.10557
7. % D&E	.81970	–.19432
6. % R&D	.81398	.05651
4. RDDE	.78604	.36563
2. INVMAC	–.09731	.89873
3. VA	.13023	.80103
5. INV	–.23012	.66510
1. SALES	.14714	.58441

Source: Author.

Horizontal factor 2, vertical factor I

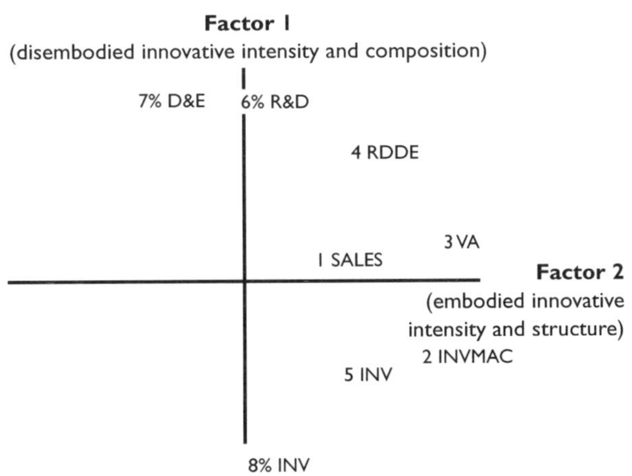

Source: Author.

Table 7 shows the results of the factor analysis. The first two factors explain a large percentage of the total variance (67.9 per cent). The rotated factor matrix and the plot of the variables according to the first two factors allow us to interpret the technological and economic meaning of the factors:

- The first factor indicates: *i)* disembodied innovative intensity measured by expenditure on R&D and D&E per employee (RDDE); and *ii)* the relevance of disembodied innovative expenditure to total innovation costs, measured as the shares of innovation costs devoted to R&D and D&E (% R&D and % D&E).
- The second factor reflects the relevance of embodied innovative activities and the features of the sectoral industrial structure represented by innovative investment (INV), the capital intensity of the sectors (INVMAC), firm size (SALES) and labour productivity (VA). This confirms the conclusions of the previous section, based on the aggregation of data at firm level, that total investment in machinery is associated primarily with the structural and organisational aspects of production, such as capital intensity, large scale, and higher labour productivity. The significant difference, however, is that, at sectoral level, the complementarity between embodied and disembodied innovative activities becomes much weaker. The link between innovative investment and total investment in machinery appears much stronger and associated (as at firm level) with firm size and labour productivity.

These results show the variety, in terms of composition and intensity of innovative activities, of sectoral innovative patterns.

Figure 1 plots 104 industrial sectors according to their position with respect to the two factors. This allows us to interpret the relationship between factor 1 (*i.e.* disembodied sources of innovation) and factor 2 (embodied sources of innovation and size-related aspects), by studying the distribution of sectors along these two dimensions.

A large number of sectors are positioned along a concave curve which shows an inverse relationship between factor 1 and factor 2. This suggests that, for most sectors, an increase in embodied innovative intensity (which is associated with size in this data set), is accompanied by a lower level of disembodied innovative intensity. However, the clustering of some sectors on the bottom left and upper right quadrant also suggests that there are quite a few exceptions to this rule. These are constituted by a group of sectors with very low innovation and a few that show both high disembodied and high embodied innovative intensity.

In order to identify the main sectors and sectoral patterns, as well as to reduce the number of sectors to manageable and meaningful categories, a factor analysis has been performed on 30 sectors. This produced the scatter plot shown in Figure 2, with the two factors representing the same variables as in Figure 1. A cluster analysis was performed on these 30 sectors to extract homogeneous groups of sectors based on the two factors.[9]

◆ Figure 1. ***Scatterplot of the sectors along the two factors (104 sectors)***

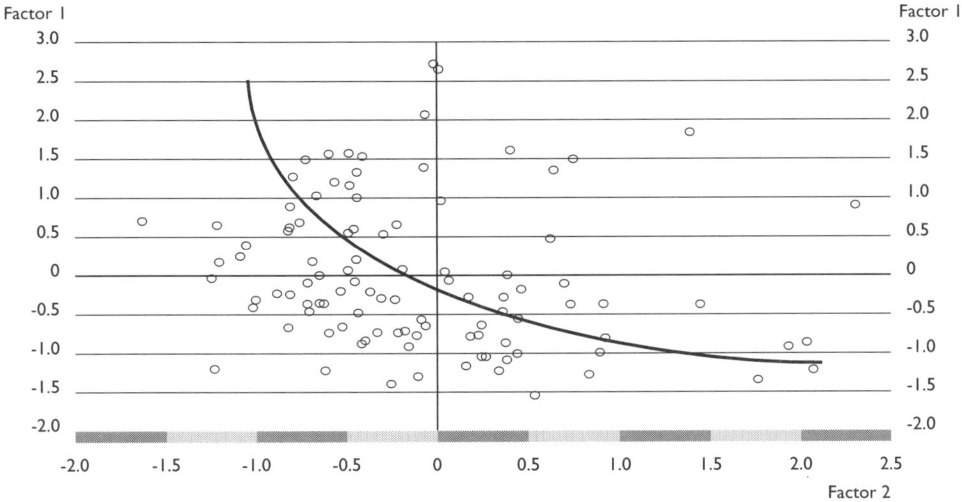

Source: Author.

◆ Figure 2. ***Disembodied and embodied sectoral innovative patterns (30 sectors)***

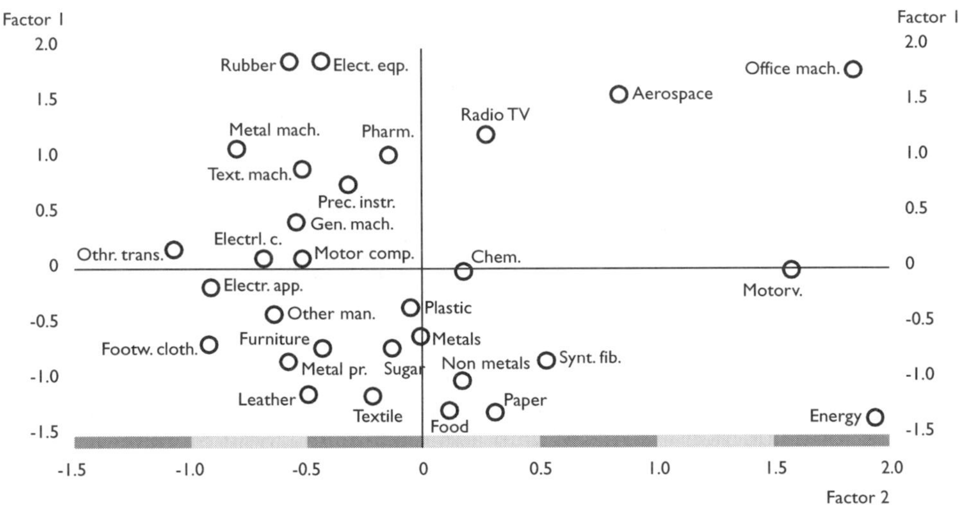

Source: Author.

The main sectoral patterns

The clustering statistical procedure was stopped at the level of 11 clusters, chosen on the basis of both the statistical and technological significance of the clusters formed. Figure 3 shows the different clusters according to their approximate position with respect to the two technological axes already used (the average values of the main innovation and structural variables used in this study are also shown). The different groups of sectors can be labelled and described as follows.

Technology users and traditional sectors

Cluster 4 is at the bottom left of Figure 3. These are sectors with low innovative intensity; they are characterised by both low disembodied and low embodied innovative intensity and a large preponderance of small firms. R&D and D&E expenditure per employee is very low, and innovative investment per employee is below average. In these sectors, firms innovate almost exclusively by purchasing and using technologies embodied in new machinery and equipment; almost 80 per cent of innovative activities rely on this source of technology. These sectors are also characterised by a low level of value added per employee, which is probably related both to the low technological content of final products and to the low capital intensity of the production processes.

Investment-intensive sectors

In this group of sectors (clusters 1-3), investment activities play an important role both as a key factor in defining the structure and organisation of production activities and as a form of innovative activity. The key technological feature of these sectors is that innovative activities are mainly embodied in investment in new machinery and equipment, while R&D and D&E play a marginal role. There are, however, significant intersectoral differences in the intensity and technological content of investment activities, as well as in the relevance of scale factors. Synthetic fibres, non-metals, food and paper and printing (cluster 2) are more based on capital intensity and scale than the sectors in cluster 3.

The combination of rather small firm size and lower investment intensity in cluster 3 compared to cluster 2 is the result of the specific characteristics of the production processes in these industries (chemicals, plastic, metals and sugar, drink, and tobacco). Also, in the case of small production units the "continuous process" nature of production activities often requires relatively capital-intensive structures of production associated with a medium-high labour productivity.

As far as innovative patterns are concerned, chemicals, plastics, metals and sugar, drink and tobacco carry out some R&D and D&E activities along with innovative investment.

Figure 3. **Disembodied and embodied innovative patterns and industrial structure**

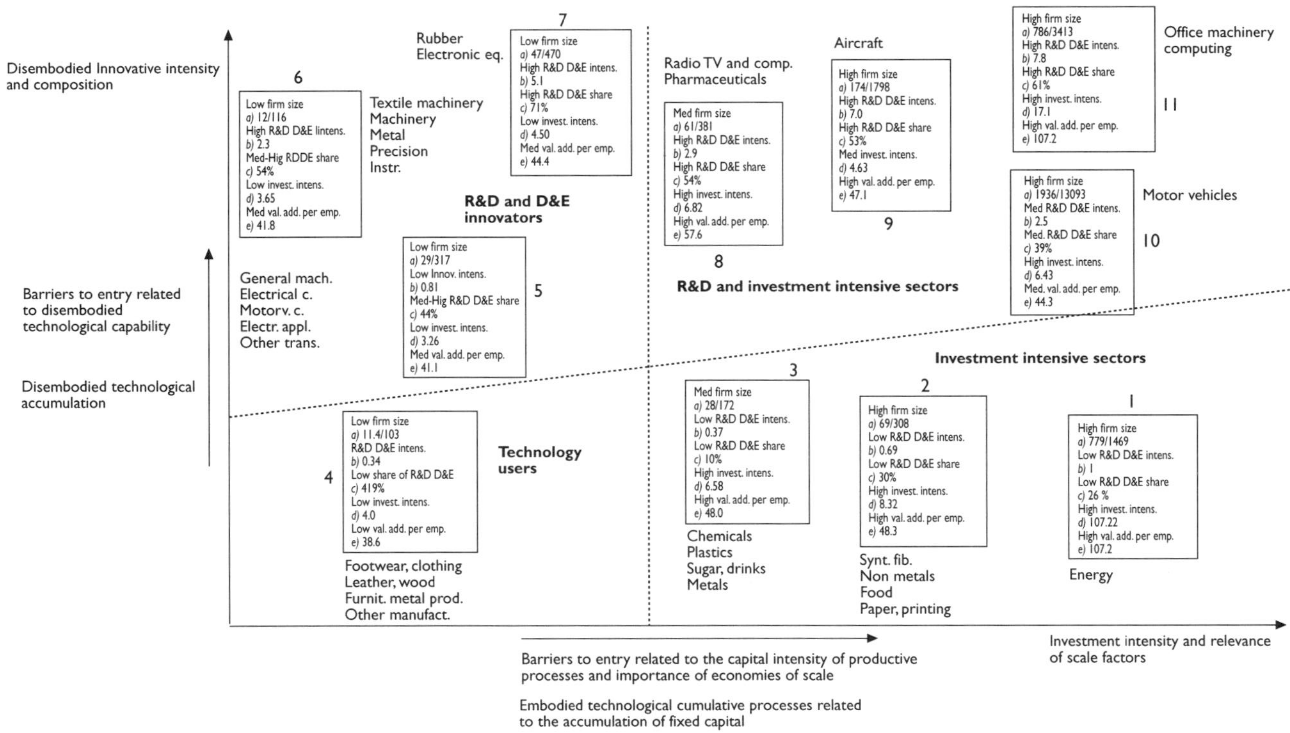

a) Sales per firm (billion lire)/number of employees per firm.
b) R&D + Engineering-Design expenditure per employee (million lire).
c) Share of R&D and D&E innovation costs.
d) Investments in machinery and equipment per employee (million lire).
e) Value added per employee (million lire).
Source: Author.

R&D and D&E innovators

Clusters 5, 6 and 7 are characterised by the combination of high expenditure on R&D and D&E per employee, small production units, and low levels of investment (both total and innovative) per employee.

Cluster 6 includes all the most innovative mechanical engineering sectors. Firm size and investment intensity are very low. In these sectors, the relatively high disembodied innovative intensity is due primarily to the relevance of D&E innovative activities. Cluster 7 shares structural characteristics with cluster 6; however, sectors in cluster 7 present higher R&D and D&E expenditure per employee. The higher innovative intensity of these sectors might also explain their relatively high value added per employee (*cf.* cluster 5).

Cluster 5, composed of some electrical and mechanical sectors, is constituted by the less innovative sectors among R&D and D&E innovators. Industrial sectors in this cluster show R&D and D&E expenditure per employee slightly above average.

R&D and investment-intensive innovators

In clusters 8, 9, 10 and 11, high R&D and D&E expenditure per employee are accompanied by a medium or high level of investment in machinery per employee and a medium or large scale of production.

Cluster 11, at the top right of the figure, concerns the office machinery and computing industry. Large firms, high expenditure on R&D and engineering activities and a surprisingly high investment intensity characterise this high-technology sector. Value added per employee is also very high and is related to the high technological content of products and processes and to the high capital intensity of these sectors.

Clusters 9 and 8 show a similar pattern, but with less extreme values, with a high disembodied innovative intensity associated with medium and large firm size and medium capital intensity. In these sectors, innovative activities aimed at producing new technological knowledge represent a large part of total innovative activities.

The motor vehicle industry (cluster 10) is an outlier of this group in terms of firm size. This reflects the fact that the automobile industry is a sector that exploits mass-production systems more extensively. Investment expenditure in machinery per employee is high, but the relevance of economies of scale is related not only to the amount of fixed capital invested but also to the organisational complexity of the productive processes. R&D and D&E intensity and their shares of total innovation costs are also above the average.

A MORE DYNAMIC INTERPRETATION OF SECTORAL INNOVATIVE PATTERNS

How can these statistical associations based on data at a given point in of time be interpreted in a somehow more dynamic perspective? One way might be to characterise the patterns identified in terms of technological barriers to entry and technological stages of industries.

The vertical and horizontal distances between clusters in Figure 3 can be considered an indicator of technological distances among sectors.[10] They can also be seen as barriers to intersectoral movements of firms and to strategies to diversify technology and production. To move from one sector to another requires overcoming structural, technological and organisational barriers to entry and to exit (due to the presence of sunk costs both in technology and production assets), the nature and importance of which vary according to the starting point and the direction of movement.[11] The vertical axis reflects the relevance of barriers to entry related to the cumulative process of disembodied and more tacit technological knowledge, whereas the horizontal axis indicates the relevance of barriers to entry related to the amount of fixed capital and accumulated embodied technology and to the complexity of the models of industrial organisation.

The double nature of the technological barriers to entry might partly explain the non-linear relationships between innovative intensity and firm size (and market structure) described in the empirical literature (Cohen and Levin, 1989).

In some industries (R&D and investment-intensive sectors in Figure 3), full exploitation of technological opportunities requires a large amount of investment, the development of complex organisations, and large volumes of production; in other innovative sectors (R&D and D&E innovators), it does not. Only in the first group does the joint presence of static and dynamic economies of scale in production and learning related to the accumulation of disembodied technological capabilities favour large firms and concentrated market structures. The relationship between firm size (and market structure) and innovative intensity therefore seems related to the extent to which disembodied and embodied technological activities are complementary innovative sources.

The different patterns in Figure 3 can also be seen in the context of the product cycle model. According to product cycle theory (Vernon, 1966; Utterback and Abernathy, 1975; Abernathy and Utterback, 1978; Utterback and Suárez, 1993), technology is fairly standardised in the mature phase of industries, and innovation is aimed at cost reduction and tends to be based on improvement in processes. This appears to be consistent with a rising proportion of embodied to disembodied technology in mature, scale-based and investment-intensive industries. New industries, instead, are associated with lower level of standardisation and may be characterised by a high proportion of disembodied to embodied technological activities.

However, the behaviour in some sectors does not fit the expectations of the product cycle approach. On the one hand, old innovative industries, such as most of the specialised machinery producers, do not seem ever to have had a large-scale standardisation stage; they continue to have many small firms, owing to the highly customised nature of their products and technologies. On the other, more recent R&D-intensive sectors, such as office machinery, telecommunications and radio-TV, have achieved, in a relatively short time, a significant level of standardisation of products, processes and technologies, together with concentrated market structures. Even our static analysis shows the complexity and interaction of the technological, structural and market dimensions that affect the dynamic of industrial sectors.

CONCLUSIONS

This discussion, based on the findings of the Italian survey, investigates the relevance of investment as a source of technology and its relationship to disembodied innovative activities at firm and sectoral level. It shows that there are large differences among sectors and firms in the importance of different sources of technology and thus confirms a broad empirical and theoretical literature (Pavitt, 1984; von Hippel, 1988; Malerba and Orsenigo, 1990; Archibugi *et al.*, 1991). These differences are not limited to the relative amount of resources devoted to innovative activities aimed at creating or improving technological knowledge (R&D and D&E); they also involve the amount of investment required to acquire and use these technologies at industrial scale, *i.e.* "embodied technological change". Innovative investment in production has emerged as a major aspect of innovation for manufacturing industry as a whole and as the most important source of innovation across firms and sectors.

Based on disaggregated data at the firm level, the analysis shows a clear complementarity between the two forms of innovative activities. However, there is also wide variance among firms in the mix of embodied and disembodied innovative activities. This suggests that sectoral differences are important explanatory factors.

At the more aggregated level of industrial sectors, the analysis shows the presence of a variety of distinct innovative patterns. Thus, in a large group of sectors, one finds a trade-off between embodied and disembodied innovative patterns, and, in a few very innovative ones, one find complementarity between R&D and investment. The clustering of industries clearly emphasizes: *i)* the predominant importance of R&D and D&E for the machinery, electrical and electronic components sectors; and *ii)* the role of investment in the chemicals, food, rubber, and metals industries; and *iii)* a combination of very high levels of R&D and investment in large-scale organisations for the most representative high-technology sectors, such as aerospace, office machinery, telecommunications, and radio-TV apparatus.

Finally, the performance indicator represented by value added per employee appears positively related across firms and sectors to greater efforts in either aspect of innovative activities, and to more structural aspects of production, such as capital intensity and firm size.

A major limitation of this analysis is the inability fully to interpret the dynamic causes underlying these results because it is not possible to analyse the changes over time. It is hoped that the forthcoming new survey data (1994) will provide some clues in this regard. Nevertheless, some explanations, in terms of accumulation of structural and technical barriers to entry and the product cycle, are suggested.

Overall, this study suggests the need, when analysing the technological profiles of firms and industries, to combine the widely used indicators of disembodied technology (such as R&D and patents) with adequate consideration of innovative investment associated with distinct forms of technological capabilities, learning processes, and organisational structures. This further suggests the need for an enlarged framework for analysing the innovative activities and growth performance of firms. Innovation surveys represent a major source of information for giving empirical support to an enlarged perspective on technological change. In this context, the recent standardisation of innovation surveys proposed by the Oslo Manual (OECD, 1992) and the ongoing European Community Innovation Survey (Eurostat, 1994) are extremely important. The latter will make available for the first time an internationally comparable database covering the key dimensions of innovation activities investigated in this article.

NOTES

1. I am grateful to Daniele Archibugi, Mario Pianta and Uwe Cantner for valuable comments. The usual disclaimer applies.
2. The survey was carried out at the level of legally defined enterprises rather than consolidated industrial groups. It was designed to provide, for individual firms, reliable quantitative evidence on a wide range of innovative activities undertaken by a large sample of Italian firms and to give additional information on the main technological sources of innovative activities, their inducement factors and economic effects. Methodological information on the CNR-ISTAT survey are reported in the Appendix. See also Cesaratto *et al.* (1991).
3. Data on innovation costs incurred by firms between 1981-85 have been divided by 5 in order to make them comparable with the industrial survey data. The matching of the two databases makes available innovation and economic data for 6 839 firms (covering more than 50 per cent of Italian manufacturing industry in terms of sales and employees).
4. In this study, the distinction between embodied and disembodied technical change does not refer to the nature of the technological output but to the nature of innovative activities, *i.e.* according to whether their goal is the production or use of technology.
5. This is because of the log-normal distribution of most of the variables considered and the non-linearity of most of the relationships. The use of logarithmic values has also allowed us to exclude firms that did not have both embodied and disembodied innovative activities.
6. The correlation coefficient computed on the total sample of firms is slightly lower and remains highly significant.
7. However, the rather low levels of the R^2 show the presence of a large intra-firm variance not explained by sectoral specificities. This has also been shown by Cesaratto and Mangano (1993) and discussed in Athreye and Evangelista (1994).
8. Corresponding to a three-digit industrial classification. Four industrial sectors with less than three firms have been excluded.
9. We used a cluster analysis based on the complete linkage method. In the cluster analyses, links are based on the distance between the different cases. There are several clustering methods, depending on how the distances are estimated. The essential aim, however, is to group the original observations in more aggregated groups in order to minimise internal variance (within each group) and maximise inter-group variance. With the complete linkage method, the distance between two clusters is calculated as the distance between their two furthest points. Other clustering methods (*i.e.* single linkage, average linkage methods) give the same results.

10. The concept of technological distance was developed by Jaffe (1986). Inter-firm technological distances have been measured by examining the patent portfolios of firms. See also Archibugi (1988) and Goto and Suzuki (1989).

11. This point has already been made by Gort and Klepper (1982, p. 634) who state that "technical change (innovations) plays a critical role in determining both entry rates and the eventual number of firms in the market" and also stress that "the number of firms in product markets technologically adjacent to those of a new product – that is, the number of potential entrants – influence the entry rates".

Annex

In October 1987-April 1988 the Italian Central Statistical Office (ISTAT), in collaboration with the Institute for Studies on Scientific Research and Documentation of the Italian National Research Council (ISRDS-CNR), carried out a survey of technological innovation in Italian manufacturing industry. The aim of the survey was to investigate the process of technological innovation and its impact on firms. The five-year period covered was 1981-85.

A preliminary survey was carried out in 1985 by means of a questionnaire mailed to about 35 000 manufacturing firms with more than 20 employees; 16 701 firms – 69.3 per cent of the 24 104 firms that answered the questionnaire – declared that they had introduced technological innovations in the period 1981-85.

On the basis of the replies to the above-mentioned preliminary survey, these firms were subdivided into two groups. One, comprising 3 200 firms, included firms that had introduced both product and process innovations based on in-house innovative activities (in particular, R&D and patents held). These firms, deemed the most innovative on the basis of the preliminary survey, were then interviewed using a 33-item questionnaire. The second group of firms, numbering about 13 000, were those having declared that they had innovated mainly by purchasing technology from outside the firm through capital goods. This second group was mailed a simplified 21-item questionnaire.

Overall, 8 220 firms responded to the questionnaire, 5 519 by post and 2 701 by direct interview.

The drop in the number of firms actually participating in the survey from the total number of 16 701 innovating firms originally identified can be explained both by the comparative complexity of the new questionnaire and by the more rigorous criteria used in the innovation survey to define technological innovation.

The 8 220 firms included in the innovation survey represent about 27 per cent of the 30 449 manufacturing companies included in the industrial survey. They account for an even higher percentage of employees, more than 52 per cent. Overall, the average size of Italian innovating firms is larger than the average size of Italian manufacturing firms.

REFERENCES

ABERNATHY, W.J. and J.M. UTTERBACK (1978), "Patterns of Industrial Innovation", *Technology Review*, No. 7, pp. 41-47.

AMENDOLA, G., G. DOSI and E. PAPAGNI (1993), "The Dynamics of International Competitiveness", *Weltwirtschaftliches*, No. 3.

ARCHIBUGI, D. (1988), "In Search of a Useful Measure of Technological Innovation", *Technological Forecasting and Social Change*, No. 34, pp. 253-257.

ARCHIBUGI, D., S. CESARATTO and G. SIRILLI (1991), "Sources of Innovative Activities and Industrial Organization in Italy", *Research Policy,* No. 20, pp. 299-313.

ARCHIBUGI, D., R. EVANGELISTA and R. SIMONETTI (forthcoming), "Concentration, Firm Size and Innovation. Evidence from Innovation Costs", *Technovation.*

ATHREYE, S. and R. EVANGELISTA (1994), "Limiting the Notion of Variety. Considerations for Empirical Research", paper presented at the 1994 EAEPE Conference, Copenhagen, 27-29 October.

CESARATTO, S., S. MANGANO and G. SIRILLI (1991), "The Innovative Behaviour of Italian Firms: A Survey on Technological Innovation and R&D", *Scientometrics*, Vol. 21, No. 1, pp. 115-141.

CESARATTO, S. and S. MANGANO (1993), "Technological Profiles and Economic Performance in the Italian Manufacturing Sector", *Economics of Innovation and New Technology*, Vol. 2, No. 2, pp. 237-256.

COHEN, W.M. and R.C. LEVIN (1989), "Empirical Studies of Innovation and Market Structure", in R. Schmalensee and R.D. Willig (eds.) *Handbook of Industrial Organization*, Vol. 2, pp. 1059-1107, Elsevier Science Publishers B.V, Amsterdam.

DOSI, G., K. PAVITT and L. SOETE (1990), *The Economics of Technical Change and International Trade*, Harvester Wheatsheaf, London.

EUROSTAT (1994), *The Community Innovation Surveys. Status and Perspectives*, Luxembourg.

FREEMAN, C. (1982), *The Economics of Industrial Innovation*, Pinter Publishers, London.

GORT, M. and S. KLEPPER (1982), "Time Paths in the Diffusion of Product Innovations", *Economic Journal*, No. 92, pp. 630-653, September.

GOTO, A. and K. SUZUKI (1989), "R&D Capital, Rate of Return on R&D Investment and Spillovers of R&D in Japanese Manufacturing Industries", *The Review of Economics and Statistics*, Vol. 71, No. 4.

von HIPPEL, E. (1988), *The Sources of Innovation*, Oxford University Press, New York.

JAFFE, A. (1986), "Technological Opportunity and Spillovers in R&D: Evidence from Firms' Patents, Profits and Market Values", *American Economic Review*, Vol. 76, No. 5.

MADDISON, A. (1991), *Dynamic Forces in Capitalist Development*, Oxford University Press, Oxford.

MALERBA, F. and L. ORSENIGO (1990), "Technological Regimes and Patterns of Innovation: A Theoretical and Empirical Investigation of the Italian Case", in A. Heertje and M. Perlman (eds.), *Evolving Technology and Market Structure*, pp. 283-385, The University of Michigan Press, Ann Arbor, Michigan.

OECD (1992), "OECD Proposed Guidelines for Collecting and Interpreting Technological Innovation Data – Oslo Manual", OECD/GD(92)26, Paris.

PAVITT, K. (1984), "Sectoral Patterns of Technological Change: Toward a Taxonomy and a Theory", *Research Policy*, No. 13, pp. 343-373.

PIANTA, M. (forthcoming), "Technology and Growth in OECD Countries, 1970-1990", *Cambridge Journal of Economics*.

SCHUMPETER, J.A. (1934), *The Theory of Economic Development* (1st edn., 1919), Harvard University Press, Cambridge, Massachusetts.

SCHUMPETER, J.A. (1939), *Business Cycles*, Mc Graw-Hill, New York.

SCHUMPETER J.A. (1942), *Capitalism, Socialism and Democracy*, Harper, New York.

SCOTT, M.F.G. (1988), *A New View of Economic Growth*, Clarendon Press, Oxford.

UTTERBACK, J.M. and W.J. ABERNATHY (1975), "A Dynamic Model of Product and Process Innovation", *Omega*, Vol. 3, No. 6, pp. 639-657.

UTTERBACK, J.M. and F.F. SUÁREZ (1993), "Innovation, Competition, and Industry Structure", *Research Policy*, No. 22, pp. 1-21.

VERNON, R. (1966), "International Investment and International Trade in the Product Cycle", *Quarterly Journal of Economics*, Vol. 80, No. 1.

INDICATORS AND DETERMINANTS OF INNOVATIVE ACTIVITY IN SWISS MANUFACTURING: AN EMPIRICAL ANALYSIS BASED ON SURVEY DATA[1]

by

Spyros Arvanitis and Heinz Hollenstein
Swiss Federal Institute of Technology, Zurich

INTRODUCTION

In earlier work, we analysed the innovative behaviour of Swiss manufacturing firms empirically, building on wide agreement in the economic literature that demand prospects, type and intensity of competition, market structure as well as factors governing the production of knowledge (appropriability, technological opportunities) are the main determinants of a firm's innovative activity (Dasgupta, 1986; Dosi, 1988; Cohen and Levin, 1989). Our analysis, based on data of the KOF/ETH Innovation Survey 1990, yielded significant parameter estimates for all four groups of variables. The explanation was quite robust across different types of innovation measures, in particular the input- and output-oriented indicators. In addition, it was concluded that – in the Swiss case – supply factors play the dominant role as determinants of innovative activity (Arvanitis *et al.*, 1992; Arvanitis and Hollenstein, 1994*a*).

The KOF/ETH Innovation Survey 1993 is part of the effort of many European countries to collect internationally comparable data on the innovative activity of firms (EUROSTAT, 1992; OECD, 1992). It produced a new database which enables us to deepen our earlier analyses. Thus, this chapter uses the new data set first to investigate the robustness of the main results of our previous study; in particular, it asks whether the empirical performance of the underlying model remains satisfactory and whether the conclusions with respect to the relative merits of supply and demand factors in explaining the innovative behaviour of firms still hold. Second, the 1993 data allows a more in-depth modelling of the appropriability of the returns to innovations using several measures (global vs. differentiated variables reflecting alternative appropriation strategies). Third, it analyses the extent to which the different external sources of knowledge contribute to the firm's own innovative activity; in this respect, the 1993 survey offers more detailed information (14 sources, as against eight in the 1990 survey). Fourth, the enlarged database for market-oriented indicators is used to explore whether this type of innovation measure, which performed rather badly in the 1990 model estimates, yields better results this time. Lastly, we present model estimates for an aggregate measure of firms' innovativeness, which has been constructed by means of a factor analysis of a whole set of input-, output- and market-oriented innovation variables.

The empirical analysis is based on the simple model of a firm that optimises its innovation output with respect to products as well as processes. It takes account of demand and market conditions, the appropriability of knowledge and technological opportunities. The neo-classical model builds primarily on the work of Dasgupta and Stiglitz (1980), Stoneman (1983), and Levin and Reiss (1988). It envisages primarily incremental innovations, *i.e.* improvements and modifications of existing key technologies or products with a certain range of basic characteristics.

In what follows, we first sketch out the underlying theoretical background and give the necessary information with respect to the specification of the different types of innovation and the explanatory variables as well as the data set used. Then we present a selection of model estimates, which may be compared to those made with the 1990 data. The final section draws some conclusions for the aforementioned topics and offers some pointers to further research.

THEORETICAL BACKGROUND AND EMPIRICAL MODEL

Theoretical background

We used as our theoretical framework a simple static deterministic model of a firm that optimises its innovation output separately for new products and new processes under monopolistic competition; the model includes the most important determinants of innovative activity found in the literature.[2] The omission of an explicit treatment of dynamic aspects and uncertainty seems to be a reasonable simplification for the theoretical setting of an empirical investigation based solely on cross-section data. The model contains the usual production and sales functions, complemented by an innovation cost and a knowledge production function with intra- and extramural knowledge as inputs; in this context the appropriability of knowledge or – viewed the other way round – know-how spillovers are taken into account. The innovation output is conceptualised as reducing cost in the case of new processes and creating demand in that of new products.

For process innovations, the demand for the firm's output q is represented by:

$$q = sp^{-e} \qquad e > 0 \tag{1}$$

The shift parameter s stands for pure income effects as well as demand influences resulting from inter-firm differences with respect to the intensity of non-price competition. The production costs c (no fixed costs, constant returns to scale) are given by:

$$c = bw^{-a}\, q \qquad 1 > a > 0 \text{ and } b > 0 \tag{2}$$

and depend – in addition to the quantity produced – on the innovation output w used as an input in the production process. The parameters b and a, together with r to be introduced below, are considered to reflect dimensions of (cost-reducing) technological opportunities. The innovation costs k for new processes (as well as for new products) are simply a linear function of the innovation output w:[3]

$$k = vw \tag{3}$$

For product innovations, the demand for the firm's output is modelled as follows:

$$q = stp^{-e} \qquad e > 0 \tag{4}$$

with:

$$t = bw^{a} \qquad 1 > a > 0 \text{ and } b > 0 \tag{5}$$

The demand schedule contains, as an additional (demand-shifting) element, the product quality t, which is improved by product innovations. Production costs are represented simply by:

$$c = c'q \qquad c' > 0 \tag{6}$$

Next, a knowledge production function is added to the model. The innovation output w is produced with two inputs: "raw" (intramural) knowledge x generated in the firm itself and "raw" (extramural) knowledge xe acquired from outside. The inputs are "raw" in the sense that they cannot be directly used for marketable new products or processes. For the present purpose, it is sufficient to assume a Cobb-Douglas-function:[4]

$$w = x^{1-r}\,(1\, x_e)^r \qquad 0 < r < 1 \text{ and } 0 < \lambda < 1 \tag{7}$$

with r as the elasticity of innovation output with respect to extramural knowledge (representing a further dimension of technological opportunity) and λ as the portion of external knowledge that can be absorbed by the firm. The degree of appropriability of the firm's own knowledge 1-λ has an impact on innovation activity by its influence on the incentive structure of the enterprise.

Profit maximisation, given equations (1), (2), (3), (7) for process and (3), (4), (5), (6), (7) for product innovations, yields first order conditions with respect to q and x, which make it possible to write x as depending on s, a, b, e, v, r and λ. A further important determinant to be considered, according to one of the Schumpeterian hypotheses, is market concentration. It is reasonable to assume in a cross-section study that the market structure, represented here by the inverse number of firms in the market 1/n, is fixed in the short run.[5]

Taking account of all elements described so far, the firm's innovative activity x is determined by the following factors:

$$x = f\,(\underset{+}{s},\ \underset{+/-}{e},\ \underset{+}{1/n},\ \underset{+}{1\text{-}\lambda},\ \underset{-/+}{b},\ \underset{+}{a},\ \underset{+}{r},\ \underset{-}{v}) \qquad (8)$$

Comparative statics yield the signs indicated in (8):[6] wherever two signs are given, the first one refers to process, the second to product innovations. The signs seem economically plausible, although there are certain problems with respect to the price elasticity of demand e, which basically result from the static character of the model. It would be more appropriate to treat e as endogenous, *i.e.* determined, among other things, by the innovative activity. In a static model, the impact of e can be rationalised by interpreting it as the elasticity anticipated by the firm. In this sense, process innovations are positively affected by price elasticity (a high e induces cost-reducing innovations); product innovations are inversely related to e (low elasticity is an incentive for the generation of new products).

A specific feature of this theoretical approach is the absence of firm size as an explanatory variable. Firm size can be used as a proxy for various economic effects connected with the innovation process, but the literature does not offer a standard economic interpretation of size effects. Therefore, it seems reasonable to concentrate on the explanatory power of variables with a clear theoretical background. An alternative research strategy would be to investigate directly hypotheses relating to different types of such size effects, but we do not do this here.[7]

Empirical model

Our approach is characterised by the parallel use of indicators related to the generation of innovations (input- and output-oriented variables) and of measures referring to market performance with new products. Moreover, an aggregate indicator for the innovativeness of a firm, based on measures from all stages of the innovation process, is included as a dependent variable. Table 1 gives some summary information on the innovation variables used in this study. As far as the first class of variables is concerned, we concentrate in this paper on one input-oriented indicator (INNIN) and one output-oriented one (IASST). Both are separately measured for product and process innovations. Estimates for these polychotomous ordered variables are based on the (ordered) probit procedure. Two types of measures may be distinguished in the market-oriented indicators: NEW and NEWGROW are related to the the early stages of a product's life cycle without reference to the degree of novelty of the products. In contrast, the degree of novelty is fundamental to the construction of alternative measures of the sales share of innovative products (SP1, SP2).[8] We employed the tobit procedure for the estimation models with these metric indicators as dependent variables. The same method of estimation is applied to the global measure of the intensity of a firm's innovation activities (GLOB). It is

Table 1. **Innovation variables**

Variable	Definition	Measurement scale/ response levels	Value range
1. Input-oriented measures			
INNIN	Global measure for input requirements based on weighted information for research, development and construction/design	Ordinal	0 to 3
2. Output-oriented measures			
IASST	Technological assessment of innovations	Ordinal 1: very low importance to 5: very high importance	0 to 5
3. Market-oriented measures			
SP1	Sales share of highly improved or new products	Metric	0 to 100
SP2	Sales share of improved, highly improved or new products	Metric	0 to 100
NEW	Sales share of products in the introduction stage of the life cycle (%)	Metric	0 to 100
NEWGROW	Sales share of products in the introduction and growth stage of the life cycle (%)	Metric	0 to 100
4. Aggregate indicator			
GLOB	Scores of factor analysis based on seven innovation measures for new products (five for processes)	Metric	Product: –2.5, 2.5 Process: –2.0, 2.0

1. The variable INNIN is constructed as follows: the values for the three input components (value range 1 to 5) are weighted with 3 (research), 2 (development) and 1 (construction/design). The weighted sum is ordered in categories (new products: <2.5, 2.5 to 3.5, >3.5; new processes: <2, 2.0 to 4.0, >4) corresponding to the response levels 1, 2 and 3 of INNIN. This variable, as well as IASST and GLOB, have been measured separately for product and process innovations, whereas market-oriented indicators represent product innovations. To include the non-innovating firms, the value range is complemented by the response level 0 for all qualitative variables as well as –2.5 (new products) and –2.0 (new processes) in the case of GLOB.

Source: KOF/ETH, *Innovation Survey 1993*; author's estimates.

developed, separately for product and process innovations, starting from the basic idea of innovativeness as an unobservable variable which has to be proxied by many indicators, each covering a specific aspect of innovative activity and prone to errors of measurement. The overall measure used in this study is constructed by means of factor analysis of a whole set of innovation indicators (some 15 for new products and about ten for new processes, of which seven and five, respectively, are retained in the final factor solution).[9]

Table 2 summarises the relevant information with respect to the four groups of explanatory variables included in the empirical model.[10] The measures used for demand expectations and market conditions (D, IPC, INPC, CONC) are self-

Table 2. **Specification of the explanatory variables**

Variable	Description/economic interpretation	Sign
1. Demand		
D	Medium-term expected change of demand	+
2. Market conditions		
IPC	Intensity of price competition in the product market: negative (positive) sign expected for product (process) innovation	–/+
INPC	Intensity of non-price competition in the product market	+
CONC	Concentration measure based on the number of principal competitors (more than 50, 16 to 50, 5 to 15, less than 5) in the world (product) market (three dummy variables)	+
3. Appropriability		
APPRG	Global measure of the extent to which innovations can be protected from competition	+
or		
APPR1, 2, 3	Scores of a factor analysis based on seven dimensions of the firm's appropriation strategy (three-factor solution): APPR1 (related to patents and design registrations) APPR2 (related to secrecy, product complexity and a lead with respect to the introduction of novelties) APPR3 (related to sales/service efforts and firm-specific human capital)	+
4. Technological opportunities		
TPOT	General technological potential, *i.e.* scientific and technological knowledge relevant to the firm' innovative activity	+
EXTINT	Specific contribution of external knowledge to the firm's own innovative activity	+
Model 2:	EXTINT is replaced by the scores of a factor analysis based on the relevance of 14 external sources of knowledge (three-factor solution):	
	SOURCE1 (related to science)	+
	SOURCE2 (related to factor inputs)	+
	SOURCE3 (related to other firms)	+

1. If not otherwise specified, the variables reflect assessments of the surveyed firms measured on a five-point scale. The variables APPRG, APPR1, 2, 3 and EXTINT as well as SOURCE1, 2, 3 have been measured separately for product and process innovations.

Source: KOF/ETH, *Innovation Survey 1993*; author's estimates.

explanatory to a large extent: D and INPC reflect different kinds of demand shifts s, PP is used as a proxy for the price elasticity of demand e, and CONC is represented by three dummy variables related to the firm-specific number n of principal competitors on the world market. This specification of CONC seems to be more appropriate for an open economy than such traditional measures as the concentration ratio or the Herfindahl index referring exclusively to the home market. The degree of appropriability of the firm's knowledge $1-\lambda$ is measured either by a global assessment variable APPRG or by three scores of a factor analysis based on the firm's assessment of the effectiveness of the elements of its strategy to preserve an innovatory lead; the best-performing factor solution yields three factors: the first refers to knowledge protection by law (APPR1), the second to secrecy, lead time and complexity of products (APPR2) and the third to sales/service efforts as well as firm-specific human capital (APPR3).[11] The technological opportunities are proxied by two (sets) of variables: the first one (TPOT) represents the general technological potential characterising the fields of activity relevant to the firm and corresponds to quality and cost shifts b. The second variable measures more specifically the contribution of external knowledge to the firm's own innovative activity (innovation elasticity of external knowledge r). This specific impact is proxied either by an overall measure (EXTINT in model 1) or a set of external knowledge sources, represented by three scores of a factor analysis computed from the firm's assessment of the relevance of 14 knowledge sources (model 2). The three factors we found refer to science (SOURCE1: private and public research institutions as well as patent information), to factor inputs (SOURCE2: acquisition of licences, investment goods and innovative firms as well as hiring of experts) and to contacts to other firms (SOURCE3: users, suppliers, competitors, firms belonging to the same conglomerate as the respondent).[12]

All variables shown in Table 2 – except CONC – reflect assessments of the surveyed firms on a five-point scale, which are introduced into the model either directly or based on a factor analysis; these assessments are assumed to be measurements on an interval scale. The sign expectations given in the last column are derived from the theoretical model; the signs for the factor scores SOURCE1,2,3 should be at least non-negative, although the sign expectations for the individual knowledge sources covered by SOURCE1,2,3 are based on *ad hoc* considerations and evidence from earlier work for the United States [for example, Cohen *et al.* (1987); Cohen and Levinthal (1989)], and for Switzerland [Arvanitis and Hollenstein (1992 and 1994*a*)].

Data

The data for the empirical tests are from the survey on innovative activity of Swiss manufacturing firms conducted by the KOF/ETH in November 1993, which was based on a stratified sample (full population of large firms, random sample for the lower two classes). About 2 500 firms were asked to fill in a questionnaire about their innovating activities in the period 1991-93. We received valid answers from 914 firms (of which 176 non-innovators) representing about 36 per cent of total employment in the Swiss manufacturing sector. The final data set includes enterprises in all fields of activity and size classes and may be considered as representative of the Swiss industry mix; however, the data show a certain bias towards larger firms. The information collected is mainly qualitative (primarily ordinal) in nature. This holds for the innovation indicators as well as the explanatory variables.

EMPIRICAL RESULTS

Estimates with indicators related to the generation of innovations

Tables 3 and 4 contain estimates of model 1 and model 2 with one input-oriented and one output-oriented qualitative innovation indicator. In model 1, three different specifications of the appropriability variable have been adopted, which allow a differentiated analysis of this dimension of spillover effects. In addition to the global assessment variable APPRG and the scores resulting from a factor analysis (APPR1,2,3), we also included in one version of model 1 size-specific variables for APPR1,2,3.[13] In model 2 we used the factor scores SOURCE1,2,3 as proxy for one of the two dimensions of technological opportunities.[14] The equations presented in Tables 3 and 4 do not include industry dummies. Estimates for all model versions that take account of industry effects do not explain more variance than those reported here; there is some interaction between the industry dummies with APPRG and INPC in the case of product innovations.

The results for both innovation variables and models as well as for product and process innovations confirm, broadly speaking, the findings of our previous work with data of the 1990 survey. All four groups of hypothesised explanatory variables seem to be statistically relevant. The strongest impact comes from the variables representing appropriability (APPRG, APPR1,2,3) and the two dimensions of the technological opportunities (TPOT as well as EXTINT and SOURCE1,2,3, but especially TPOT). Whereas the medium-term demand expectations D are also highly significant, the impact of the variables describing the structural conditions on the product market is rather weak and not very robust; clear effects result only for price and non-price competition (IPC, INPC), whereas the market concentration dummies (CONC) are never significant. On the whole, and in accordance with previous results, the supply-oriented determinants of innovations seem to be more relevant than the variables to be attributed to a primarily demand-oriented explanation.

A more detailed inspection of the results of Tables 3 and 4 leads to further interesting findings. First, the impact of appropriability is quite similar with both measures for both innovation indicators. However, the estimations with the differentiated variable APPR1,2,3 reveal a notable difference between product and

Table 3. **Ordered probit estimates with INNIN (models 1 and 2)**[1]

Explanatory variable	Product innovations				Process innovations			
	1	2	3	4	5	6	7	8
D	.15***	.15***	.15***	.15***	.14***	.14**	.14**	.14***
	(.05)	(.06)	(.06)	(.05)	(.05)	(.06)	(.06)	(.06)
IPC	.01	.00	.01	.03	.00	.03	.03	.03
	(.05)	(.05)	(05)	(.05)	(.05)	(.05)	(.05)	(.05)
INPC	.08*	.10**	.09**	.10**	.02	.04	.04	.04
	(.04)	(.05)	(.05)	(.04)	(.04)	(.05)	(.05)	(.05)
CONC								
6-15 firms	–.02	–.07	–.08	.04	–.07	–.12	–.12	.02
	(.13)	(.14)	(.14)	(.13)	(.13)	(.14)	(.14)	(.14)
16-50 firms	.03	.00	–.01	.05	–.01	–.09	–.09	.03
	(.15)	(.16)	(.16)	(.15)	(.15)	(.16)	(.17)	(.16)
> 50 firms	–.12	–.18	–.20	–.11	–.11	–.14	–.16	–.16
	(.16)	(.17)	(.17)	(.16)	(.16)	(.17)	(.17)	(.17)
APPRG	.10**			.08*	.14***			.11**
	(05)			(.03)	(.05)			(.05)
APPR1		.12***				.06		
		(.05)				(.05)		
Small firms			.08				–.01	
			(.06)				(.07)	
Big firms			.20**				.19**	
			(.08)				(.09)	
APPR2		.23***				.25***		
		(.06)				(.06)		
Small firms			.26***				.23***	
			(.07)				(08)	
Big firms			.18				.37***	
			(.12)				(.12)	
APPR3		.08				–.03		
		(.06)				(.06)		
Small firms			–.11				–.06	
			(.07)				(.08)	
Big firms			–.02				.04	
			(.10)				(.11)	
TPOT	.18***	.15***	.14***	.18***	.20***	.18***	.18***	.18***
	(.04)	(.05)	(05)	(.05)	(.04)	(.05)	(.05)	(.05
EXTINT	.13***	.11**	.11**		.08*	.08*	.09*	
	(.05)	(.05)	(.05)		(.05)	(.05)	(.05)	
SOURCE1				.10*				.18***
				(.06)				(.05)
SOURCE2				.04				.11*
				(.06)				(.06)
SOURCE3				.02				–.03
				(.06)				(.06)
N	622	557	557	601	619	541	541	596
McFadden R^2	0.037	0.045	0.047	0.035	0.040	0.047	0.051	0.049
Schwartz Crit. SC	1689	1517	1534	1650	1447	1252	1266	1380
Somers' D	0.260	0.284	0.291	0.260	0.288	0.318	0.326	0.326
ESA (Prob.)	(.08)	(.12)	(.22)	(.02)	(.00)	(.00)	(.00)	(.00)

1. Each column includes the estimated parameters and the standard errors in brackets; the statistical significance is indicated with ***, **, and * representing the 1 per cent, 5 per cent and 10 per cent levels, respectively. Intercepts have been omitted throughout. ESA-Test: Score Test for Equal Slopes Assumption.

Source: Author's estimates.

Table 4. **Ordered probit estimates with IASST (models 1 and 2)**

Explanatory variable	Product innovations			Process innovations				
	1	2	3	4	5	6	7	8
D	.12**	.13**	.13**	.09*	.13***	.12**	.12**	.11**
	(.05)	(.05)	(.06)	(.05)	(.05)	(.06)	(.06)	(.05)
IPC	.00	.00	.01	.01	.10**	.10**	.11**	.11**
	(.04)	(.05)	(.05)	(.05)	(.05)	(.05)	.05)	(.05)
INPC	.08*	.07	.07	.10**	.07*	.09*	.09**	.10**
	(.04)	(.05)	(.05)	(.04)	(.04)	(.05)	(.05)	(.04)
CONC								
6-15 firms	.16	.14	.10	.14	–.00	–.08	–.09	.00
	(.13)	(.14)	(.14)	(.13)	(.13)	(.14)	(.14)	(.13)
16-50 firms	.09	.03	–.02	.08	–.05	–.13	–.16	–.06
	(.15)	(.16)	(.16)	(.15)	(.15)	(.16)	(.16)	(.15)
> 50 firms	.14	.13	.09	.05	–.05	–.08	–.10	–.20
	(.15)	(.17)	(.17)	(.16)	(.15)	(.16)	(.17)	(.16)
APPRG	.14***			.15***	.06			.06
	(.05)			(.05)	(.04)			(.05)
APPR1		.10**				–.03		
		(.05)				(.05)		
Small firms			.01				–.10	
			(.06)				(.07)	
Big firms			.27***				.10	
			(.08)				(.09)	
APPR2		.24***				.20***		
		(.06)				(.06)		
Small firms			.25***				.21***	
			(.07)				(.08)	
Big firms			.24**				23**	
			(.11)				(.11)	
APPR3		–.02				–.05		
		(.05)				(.06)		
Small firms			–.01				–.07	
			(.06)				(.07)	
Big firms			–.00				.04	
			(.10)				(.10)	
TPOT	.21***	.18***	.17***	.22***	.19***	.19***	.19***	.20***
	(.04)	(.05)	(.05)	(.04)	(.04)	(.05)	(.05)	(.05)
EXTINT	.12***	.10**	.10**		.19***	.17***	.17***	
	(04)	(.05)	(.05)		(.04)	(.05)	(.05)	
SOURCE1				.01				.09*
				(.05)				(.05)
SOURCE2				–.04				–.13**
				(.06)				(.06)
SOURCE3				.08				.15***
				(.06)				(.05)
N	630	554	554	607	621	534	534	594
McFadden R^2	0.038	0.039	0.043	0.032	0.038	0.043	0.045	0.036
Schwartz Crit. SC	2024	1782	1794	1965	1959	1686	1701	1888
Somers' D	0.283	0.276	0.293	0.255	0.283	0.302	0.313	0.263
ESA (Prob.)	(.02)	(.06)	(.00)	(.00)	(.00)	(.00)	(.00)	(.00)

1. Each column includes the estimated parameters and the standard errors in brackets; the statistical significance is indicated with ***, **, and * representing the 1 per cent, 5 per cent and 10 per cent levels, respectively. Intercepts have been omitted throughout. ESA-Test: Score Test for Equal Slopes Assumption

Source: Author's estimates.

process innovations: patents and other institutional mechanisms of knowledge protection are of importance only for new products, whereas secrecy, lead time and product complexity are influential methods of protection for both kinds of innovation. On the other hand, sales and service efforts do not seem to have any significant effect on innovative activity. Further, the results show that patents (and other institutional means) are more suitable instruments for the larger firms; small and big firms alike use, apparently in a rather successful way, other protection strategies (APPR2).

Second, with respect to technological opportunities, the contribution of technological potential proves to be stronger than that of external knowledge in both model versions, a result opposite to that obtained with the 1990 data. SOURCE1, comprising information from private and public research institutions (including universities) as well as information disclosed by patent specifications, is important for both kinds of innovation but more pronounced for the input-oriented indicator, which measures the intensity of R&D inputs. Information coming from other firms only has a signicant effect for process innovations if measured by an output-oriented indicator, once more a departure from the results of our previous work. Surprisingly, only a slight contribution comes from sources such as licences, investment goods and exports or the acquisition of innovative firms (we have so far no explanation for the negative sign of this variable in column 8 of Table 4).

Third, some remarks should be made concerning the impact of market variables. The intensity of non-price competition yields the expected positive sign for both types of innovation. A positive impact is also reported for the intensity of price competition in the case of process innovations measured by IASST. The somewhat worse performance of this variable with the 1993 data is probably due to business cycle effects that increase the intensity of price competition for all industries, independently of firms' innovation efforts. Concentration in the product market has no significant impact on innovation behaviour. There is no appreciable correlation between the appropriability variables and market concentration; additional model estimates not reported here show that the impact of the appropriability variables is not influenced by market concentration, a hypothesis which is sometimes postulated in the literature (Spence, 1984). Fourth and last, favourable demand prospects seem to stimulate innovations in boom as well as in recession phases of the business cycle.

Estimates with market-oriented measures of innovation

Table 5 shows model 1 and model 2 estimates for the two types of market-oriented innovation indicators used in this study. An overall look at the results points to a better model fit for NEWGROW than for the other three variables, possibly because it may be more difficult for respondents to attach their products

Table 5. **Tobit estimates with market-oriented innovation measures (models 1 and 2)**

Explanatory variable	Product innovations							
	Model 1				Model 2			
	SP1 1	SP2 2	NEW 3	NEWGROW 4	SP1 5	SP2 6	NEW 7	NEWGROW 8
D	4.55***	6.65**	2.01***	8.26***	3.28**	5.33***	1.43**	7.69***
	(1.51)	(1.98)	(0.73)	(1.45)	(1.46)	(1.87)	(0.68)	(1.38)
IPC	0.36	0.30	–0.04	–0.91	–0.56	0.48	–0.67	–1.74
	(1.31)	(1.73)	(0.65)	(1.29)	(1.29)	(1.65)	(0.62)	(1.24)
INPC	1.54	2.39	0.08	1.19	1.87	3.14**	0.14	0.95
	(1.24)	(1.64)	(0.61)	(1.21)	(1.23)	(1.58)	(0.57)	(1.15)
CONC								
16-50 firms	1.20	0.11	1.43	7.64*	5.32	4.22	1.89	8.63**
	(4.29)	(5.62)	(2.06)	(1.45)	(4.32)	(5.47)	(1.97)	(3.97)
6-15 firms	5.83	6.52	1.35	8.37**	8.55**	8.63*	0.62	7.55**
	(3.73)	(4.88)	(1.80)	(3.56)	(3.72)	(4.69)	(1.69)	(3.40)
< 5 firms	3.15	7.02	–1.21	1.42	5.17	7.88	–2.00	–0.15
	(4.63)	(6.08)	(2.28)	(4.48)	(4.59)	(5.82)	(2.13)	(4.26)
APPRG					0.36	0.40	1.04	2.66**
					(1.37)	(1.78)	(0.65)	(1.31)
APPR1	–0.21	0.73	0.68	0.28				
	(1.29)	(1.70)	(0.62)	(1.23)				
APPR2	4.99***	7.70***	2.74***	5.94***				
	(1.72)	(2.28)	(0.86)	(1.67)				
APPR3	2.52	2.48	1.47*	2.56*				
	(1.55)	(2.05)	(0.76)	(1.50)				
TPOT	0.30	3.03*	1.31**	1.97*	2.03*	4.51***	1.42**	2.48**
	(1.26)	(1.67)	(0.62)	(1.22)	(1.25)	(1.61)	(0.58)	(1.17)
EXTINT	1.29	1.15	0.80	1.05				
	(1.31)	(1.73)	(0.65)	(1.29)				
SOURCE1					–1.09	1.26	–0.14	0.76
					(1.50)	(1.95)	(0.71)	(1.43)
SOURCE2					0.17	0.88	0.87	1.68
					(1.65)	(2.14)	(0.78)	(1.57)
SOURCE3					6.39***	4.15*	1.54**	1.25
					(1.69)	(2.15)	(0.79)	(1.58)
N	505	505	498	498	557	557	554	554
Left censored	176	110	129	67	197	118	145	79
Right censored	3	18	0	7	4	22	0	7
LR-test	33.6***	46.0***	46.8***	77.3***	35.8***	36.6***	32.4***	68.0***

1. Each column includes the estimated parameters and the standard errors in brackets; the statistical significance is indicated with ***, **, and * representing the 1 per cent, 5 per cent and 10 per cent levels, respectively. Intercepts have been omitted throughout.

Source: Author's estimates.

unequivocally to several degrees of novelty (see SP1 and SP2 in Table 1) or to define the introductory stage of the product cycle (NEW) than to measure sales for the sum of the introductory and the growth stage (NEWGROW).

Although the model fit is not impressive, the estimates correspond better to the underlying model than those obtained with the 1990 data.[15] However, the results are worse than those for input- and output-oriented indicators. The main differences are related to the supply-side variables, which are less important for market-oriented measures of innovation. This holds primarily for technological opportunities: the significance of the technological potential is markedly reduced or even missing, and there is no impact from the overall measure of external knowledge. Looking in more detail at the results, the pattern of explanation shows three (economically plausible) differences compared to the indicators related to the generation of innovations. First, among the knowledge sources, "other firms" are most important, whereas "science" was dominant in the case of input-oriented measures. Secondly, with respect to appropriability, APPR1 (legal protection) is very important for input-and output-oriented measures and APPR3 (protection mechanisms related to sales and service efforts) for market-oriented variables; in both cases APPR2 is a highly significant source of protection.[16] Thirdly, there is some role for concentration in explaining market-oriented measures of innovation; we find – although not in all equations – a positive impact from an intermediate degree of market concentration, whereas polypolistic markets or a very high degree of concentration (below five competitors) do not foster innovative activity.

In sum, the results obtained for market-oriented innovation measures are not up to the expectations expressed by OECD (1992), Eurostat (1992), or Kleinknecht (1993). One possible problem with this type of innovation variable may be the blurred definition of the different categories of new products; therefore, before the next round of innovation surveys, it would be advisable to test carefully alternative wordings of this type of question. It may also be that the relationship between a (technical) innovation and success in the market place is not straightforward. Because these indicators seem to catch – at least partly – effects of market creation and development, they cannot be explained exclusively by the determinants of innovative activity; an adequate model has probably to include additional variables which reflect the firm's marketing capabilities (quality of its distribution network, advertising activities, image with users, etc.). Thus, market-oriented measures would have to be distinguished from input-and output-oriented indicators related to the generation of novelties. This view is supported by the results of a factor analysis of the main innovation indicators, which, in a solution with two factors, clearly separates market-oriented indicators from those related to the generation of innovations (Hollenstein, 1995).

Table 6. **Tobit estimates with an aggregate measure of innovativeness (models 1 and 2)**

Explanatory variable	Product innovations			Process innovations		
	All firms		Innovators	All firms		Innovators
	1	2	3	4	5	6
D	.27***	.26***	.08*	.15**	.16**	.06
	(.08)	(.08)	(.04)	(.07)	(.07)	(.05)
IPC	.05	.05		.06	.07	.06
	(.07)	(.07)	(.04)	(.06)	(.06)	(.04)
INPC	.15**	.15**	.12***	.10*	.10*	.09**
	(.07)	(.07)	(.04)	(.05)	(.05)	(.04)
CONC						
16-50 firms	.17	.17		−.02	.05	−.09
	(.24)	(.24)	(.14)	(.18)	(.18)	(.13)
5-15 firms	.30	.27	.04	.11	.15	.00
	(.21)	(.21)	(.12)	(.16)	(.16)	(.11)
1-4 firms	.25	.26	−.03	. 14	.12	−.06
	(.25)	(.25)	(.14)	(.20)	(.20)	(.14)
APPR1	.14**	.16**	.10**	.07	.04	.11**
	(.07)	(.07)	(.04)	(.06)	(.06)	(.04)
APPR2	.42***	.41***	.24***	.35***	.34***	.26***
	(.10)	(.10)	(.06)	(.07)	(.07)	(.05)
APPR3	−.01	−.03	.17***	−.04	−.05	.19***
	(.09)	(.09)	(.05)	(.07)	(.07)	(.05)
TPOT	.23***	.27***	.13***	.23***	.21***	.13***
	(.07)	(.07)	(.04)	(.05)	(.05)	(.04)
EXTINT	.09			.07		
	(07)			(.05)		
SOURCE1		−.07	−.01		−.01	−.09*
		(.08)	(.05)		(.07)	(.05)
SOURCE2		−.05	.09*		.09	.17***
		(.09)	(.05)		(.07)	(.05)
SOURCE3		.15	.07		.13**	.10**
		(.10)	(.06)		(.06)	(.04)
N	437	437	347	494	494	407
Censored values	90	90		87	87	0
LR-test	71.2***	74.6***	92.4***	70.4***	72.6***	116.6***

1. Each column includes the estimated parameters and the standard errors in brackets; intercepts have been omitted throughout. The statistical significance of the parameter estimates and the LR-test is indicated with ***, **, and * representing the 1 per cent, 5 per cent and 10 per cent levels, respectively.

Source: Author's estimates.

Estimates with an aggregate measure of innovativeness

Table 6 shows estimates of model 1 and model 2 for an aggregate measure of innovativeness, which is based on indicators from all stages of the innovation process. The results for all firms (*i.e.* non-innovators included) confirm the robustness of the underlying model. Again, all four groups of variables play a significant role. Compared to estimates with variables related to the generation of innovations (input- and output-oriented indicators), the main difference is a shift among the two variables standing for technological opportunities towards the (general) technological potential; the very weak impact of (more firm-oriented) external knowledge is in line with the results for market-oriented innovation indicators.

The regressions for the sub-sample of innovators show a better overall model fit than the estimates for the full sample. The relative importance of supply-oriented explanatory variables is even more pronounced in this case; appropriability weighs more, in view of the significance of all three appropriation factors, whereas the impact of demand expectations is weakened. In sum, the results show that the model not only explains the behaviour of innovators as it contrasts to that of non-innovating firms but also differences with respect to the intensity of innovative activity.

CONCLUSIONS

The present study, based on a new data set (KOF/ETH Innovation Survey 1993), confirms to a large extent the results of similar work with data from the previous survey (1990). Again, we find a pattern of explanation which is robust across input- and output-oriented innovation indicators, whereas the results for market-oriented measures are somewhat less convincing. The four groups of explanatory variables – *i.e.* demand prospects, market conditions, appropriability and technological opportunities – that are the ingredients of the underlying theoretical framework, contribute all, but in different proportions, to an explanation of the innovative activity of firms.

As far as the relative importance of supply and demand factors is concerned, the clear dominance of supply-oriented variables (appropriability, technological opportunities), which we found in previous work, is confirmed (Table 7). This result holds for product as well as process innovations. In both cases – it is even more pronounced for new processes – technological opportunities have the greatest explanatory power of the four groups of variables defined in the model. The variables related to market conditions again prove the least important. In sum, the evidence for Switzerland is again more in line with the "technology-push" than with the "demand-pull" hypothesis of innovation.

Table 7. **Breakdown of the explained variance by group of variables**

As a percentage of explained variance

Explanatory variable	New products				New processes			
	Input measures		Output measures		Input measures		Output measures	
	1	2	1	2	1	2	1	2
Demand	13	12	7	6	12	10	6	6
Market conditions	5	10	5	7	0	2	10	13
Appropriability	25	27	26	30	27	12	8	11
Technological opportunities	57	51	62	57	61	76	76	70

1. The figures for output-oriented estimates are based on the equations for IASST, for input-oriented measures on INNIN (unweighted mean of the percentages if more than one equation is available). Equivalent information from tobit estimates (SP, SPW, NEW, NEWGROW) is not available for technical reasons. Both model versions, indicated by 1 and 2, are used in the calculations for each type of measurement.

Source: Author's estimates.

With respect to a more in-depth modelling of appropriability and the contribution of external knowledge sources, the results may be summarised as follows: more differentiated measures of the appropriability of knowledge are superior to less specific ones. The more detailed information available for external knowledge sources did not lead to better results than the work based on the 1990 data; on the contrary, an overall measure for the contribution of external knowledge to the firm's own innovative activity proved this time – at least for input- and output-oriented measures of product innovations – to be superior to more differentiating modelling.

The estimates for market-oriented measures, even the best ones, do not match those for the other types of innovation indicators, but are clearly better than those using the 1990 data. This somewhat disappointing result may partly reflect conceptual problems, but probably indicates, first of all, the necessity to complement the underlying model with variables pertaining to the firm's marketing capabilities. Thus, a clear distinction is required between input and output indicators related to the generation of novelties on the one side, and market-oriented variables on the other side.

The results for an aggregate measure of innovativeness confirm the general robustness of the underlying model. In addition, according to estimates for the sub-sample of innovators, the model not only explains the behaviour of innovators as opposed to non-innovating firms but also differences with respect to the intensity of innovative activity.

In the light of these results, we propose two priorities for further research. First, it would be interesting to see whether the results of this paper are corroborated by corresponding studies for other countries disposing of similar data sets. Up to now, data limitations seriously impeded such international comparisons.[17] Preliminary work by König *et al.* (1994) with German data seems to confirm the not-so-convincing results we presented for the market-oriented variables SP1 and SP2. Second, it is necessary to analyse in more detail the relationship between innovation and the success of firms in the market place, applying a whole range of measures: market-oriented innovation indicators, such as those used here, as well as more general measures of a firm's market performance based on data for sales, value added or profits. Such work could bring together the large body of literature pertaining to the relationship between knowledge capital, spillovers and productivity and investigations like the present one focused on the analysis of the innovation process in a more narrow sense.

NOTES

1. The research reported in this paper was supported by the Federal Ministry of Economic Affairs of Switzerland.
2. See Arvanitis (1991) for a detailed description of the model.
3. Assuming increasing marginal costs does not change the basic results given below.
4. For the implications of other functional forms, see Arvanitis (1991).
5. It is possible to include concentration in a formal model of the type discussed above, but for slightly different market conditions, *i.e.* under the assumption of a Cournot oligopoly, it can be shown, that the results obtained so far remain valid also for that case.
6. The character of the model only allows establishing the sign and not the magnitude of the individual effects.
7. Regression estimates with firm size as an additional explanatory variable not presented here show that it it has an (additional) impact on innovation; hence, it catches some relevant but unspecific explanatory factors. However, there is also sufficient correlation between firm size and some of the explanatory variables to support our suspicion that firm size is a non-specific proxy for too many things. For a detailed analysis of the impact of firm size, see Arvanitis (1995).
8. Estimates for other indicators not presented in this paper are, on the input side, quantitative measures for innovation expenditures in the narrow (R&D) or broad sense (inclusion of expenditures for contruction/design or even follow-up investments such as production and marketing preparation). Among the output indicators, the firm's assessment of its novelties in economic terms and the number of patents and registrations of design are not considered in this study. The same holds for estimates that include the sales share of world novelties as an additional market-oriented indicator.
9. A somewhat more complex factor solution contains information from eight (new products) and seven (new processes) indicators. See Arvanitis and Hollenstein (1994*b*) and Hollenstein (1995) for a detailed description of the procedure used, the computation and interpretation of solutions with two factors (not considered here) and for empirical tests related to the quality of an aggregate measure of innovativeness in statistical and economic terms.
10. No measures are available in our data set for a, the elasticity of costs (demand) with respect to innovation output, and for v, the unit costs of innovation.
11. Our earlier work – failing more specific information – relied exclusively on APPRG.
12. The KOF/ETH Innovations Survey 1990 yielded information for only eight external knowledge sources, which were all included in the estimation of model 2.

13. We considered only two size classes: firms with more than 200 employees and firms with less than 200 employees. The threshold was put at 200 because, first, it corresponds to experience from other studies of the Swiss manufacturing sector and, second, it enables us to have enough cases for valid estimations with both classes.

14. Explorative estimations with different sets of the measured external knowledge sources yielded unsatifactory results due to relatively strong correlation among these variables.

15. Estimates with the 1990 data yielded a significant impact on NEW only for APPRG and the knowledge source USER; with NEWGROW we got a positive sign for D and a negative one for IPC (Arvanitis and Hollenstein 1994*a*).

16. The superiority of a differentiated modelling of appropriability – APPR1,2,3 vs. APPRG – is more marked for market-oriented innovation indicators than for the other two types.

17. For comparisons with the results of some studies for the United States, see Arvanitis and Hollenstein (1994*a*); more detailed information on such comparisons is found in Arvanitis *et al.* (1992).

REFERENCES

ARVANITIS, S. (1991), "Bestimmungsfaktoren des Innovationsverhaltens der Unternehmung: Modelltheoretische Grundlagen zur Herleitung eines empirischen Ansatzes", *KOF/ETH-Arbeitspapier*, No. 38, Zurich, November.

ARVANITIS, S. (1995), "Innovation and Firm Size – An Empirical Investigation Based on Swiss Firm Data", paper presented at the 39th International Atlantic Economic Conference, Vienna, March 11-15.

ARVANITIS, S. and H. HOLLENSTEIN (1992), "The Determinants of Innovative Activity of Swiss Manufacturing", in K. H. Oppenländer and G. Poser (eds.), *Business Cycle Analysis by Means of Economic Surveys,* papers presented at the 20th CIRET Conference, Budapest 1991, Vol. 2, pp. 115-146, Avebury, Aldershot.

ARVANITIS, S. and H. HOLLENSTEIN (1994*a*), "Demand and Supply Factors in Explaining the Innovative Activity of Swiss Manufacturing Firms", *Economics of Innovation and New Technology*, Vol. 3, pp. 15-30.

ARVANITIS, S. and H. HOLLENSTEIN (1994*b*), "Die Messung der Innovationsintensität: Eine empirische Untersuchung anhand schweizerischer Unternehmensdaten", *Schweiz. Zeitschrift für Volkswirtschaft und Statistik*, Vol. 130, Heft 3, pp. 523-537.

ARVANITIS, S., R. ETTER, A. FRICK and H. HOLLENSTEIN (1992), "Innovationsfähigkeit und Innovationsverhalten der Schweizer Wirtschaft. Bestandesaufnahme und Ueberprüfung von Hypothesen auf der Basis von Firmendaten", *Studienreihe Strukturberichterstattung,* Bundesamt für Konjunkturfragen, Bern-Zurich.

COHEN, W.M. and R.C. LEVIN (1989) "Empirical Studies of Innovation and Market Structure", in R. Schmalensee and R.D. Willig (eds.), *Handbook of Industrial Organization*, North-Holland, London.

COHEN, W.M., R.C. LEVIN and D.C. MOWERY (1987), "Firm Size and R&D Intensity: A Reexamination", *Journal of Industrial Economics*, Vol. 35, pp. 543-565.

COHEN, W.M. and D.A. LEVINTHAL (1989), "Innovation and Learning: The Two Faces of R&D" *Economic Journal,* Vol. 99, pp. 569-596.

DASGUPTA, P. (1986), "The Theory of Technological Competition", in J.E. Stiglitz and G.F. Mathewson (eds.), *New Developments in the Analysis of Market Structure*, MIT Press, Cambridge, MA.

DASGUPTA, P. and J.E. STIGLITZ (1980), "Industrial Structure and the Nature of Economic Activity", *Economic Journal*, Vol. 90, pp. 266-293.

DOSI, G. (1988), "Sources, Procedures, and Microeconomic Effects of Innovation", *Journal of Economic Literature*, Vol.26, pp. 1120-1171.

EUROSTAT (1992), "Proposed Questions for the Harmonized Innovations Surveys", Final Version, Luxemburg, July.

HOLLENSTEIN, H. (1995), "An Aggregate Indicator for the Innovativeness of a Firm – An Empirical Analysis Based on Survey Data", paper presented at the 39th International Atlantic Economic Conference, Vienna, March 11-15.

KLEINKNECHT, A. (1993), "Testing Innovation Indicators for Panel Surveys: Results from a Five Country Project", in A. Kleinknecht and D. Bain (eds.), *New Concepts in Innovation Output Measurement*, St. Martins Press, London, pp. 153-188.

KOENIG, H., G. LICHT and M. STAAT (1994), "F&E-Kooperationen und Innovationsaktivität", Zentrum für Europäische Wirtschaftsforschung und Universität Mannheim, mimeo, Mannheim, May.

LEVIN, R.C. and P.C. REISS (1988), "Cost-reducing and Demand-creating R&D with Spillovers", *Rand Journal of Economics*, Vol. 19, pp. 538-556.

OECD (1992), "OECD Proposed Guidelines for Collecting and Interpreting Technological Innovation Data – Oslo Manual", OCDE/GD(92)26, Paris.

SPENCE, M. (1984), "Cost Reduction, Competition, and Industry Performance", *Econometrica*, Vol. 25, pp. 101-121.

STONEMAN, P. (1983), *The Economic Analysis of Technological Change*, Oxford University Press, Oxford.

8

THE LOCALISATION OF CLUSTERS OF INNOVATIVE ACTIVITY IN ITALY, FRANCE AND CHINA[1]

by
Chris DeBresson
School of Management Sciences, University of Quebec, Montreal
and
Hu Xiaoping
McGill University, Montreal

INTRODUCTION

Using recent statistical surveys of innovative activity, we compile or estimate "innovative matrices" and compare them with Leontief input-output tables. The three countries studied share certain features: innovative activities form clusters in certain parts of their national economies, and this occurs where industries have the greatest number of downstream and upstream domestic economic linkages. These clusters of innovative activity are found, however, in different sectors of the national economy, thereby emphasizing national differences in inter-industrial innovative systems.

Because technological innovation has become a key aspect of international competition, every country is interested in knowing where its technological advantages lie. Yet, previous science and technology indicators are not very helpful in this respect. In all three countries studied, the pharmaceutical industry and the aircraft industry are the two most R&D-intensive industries; at the other end of the spectrum, the leather industry, everywhere, does very little R&D. In every country, the pharmaceutical, chemical and mechanical industries have the greatest propensity to patent inventions; at the other end of the spectrum, the electronics industry seeks few patents. Since industries behave similarly in all countries, it is difficult to identify national specialisations by using R&D expenditure and patents. While close analysis of the data may reveal some specialisation ratios, their interpretation is not as clear-cut as decision makers would wish. Recent surveys of innovative activities provide a better tool for highlighting country specialisations.

Before showing why this is so, we should explain why we use the broad term "innovative activities" and not the more restrictive "technological innovation". The reason is simple. A major innovation, as the source of growth and development, can only be identified long after its introduction. Without transistors, integrated circuits, higher level programming languages, and microprocessors, for example, use of the digital computer would have remained confined to science, defence and planning agencies. Therefore, in 1958, Jewkes was right not to name digital computing as a major post-war innovation in the first edition of his famous book, *The Sources of Invention*. Then, in the second edition, after the transistor and integrated circuits had had their impact in the 1960s, he included it. In fact, the success of an innovative endeavour depends on the subsequent emergence of complements and applications. Whether or not an innovation will be adopted is very hard to predict.

Industrial surveys cannot identify innovation, even of a minor type, in a timely fashion. What these surveys can identify in a timely fashion, however, is a broad web of innovative efforts, among which there are, as we discover much later, those rare pearls: major innovations.

Indicators of innovative activity have two major advantages over R&D and patent indicators. First, although major innovations are very rare, businesses that engage in some form of innovative activity are far more numerous than those that patent or engage in R&D. In all surveys – whether population or sample surveys – between 40 and 50 per cent of businesses are estimated to engage in some form of innovative activity. Only about 10 per cent of these engage in R&D or obtain patents. Technologically, these innovative activities may often be of minor importance: adopting new machinery or a new material, improving a product or process, adapting these to a new application, or combining existing techniques in a new way. There is, however, little fundamental difference between the type of activity involved in minor and major innovations, and the innovator cannot predict either the ultimate private benefits, or, *a fortiori,* the social ones.What may begin as an effort to obtain a simple improvement in performance may result in a major innovation. There are therefore sound analytical grounds for considering all these activities together. The large number of observations also provides an added advantage: it helps make it possible to differentiate among country performances.

Second, innovative activities reflect use of technology,[2] which differentiates among countries better than supply of technology. R&D is an input indicator which tells nothing about the use to which it is put; in fact, much is never used. Patented inventions are an intermediate output of technical activity; but most are never used. And, although most patent applicants have to demonstrate the potential usefulness of their inventions, only Canada asks the applicant to identify the potential use of the invention, which is then coded by the patent office. However, patent indicators tell us little about specialisation in the use of the techniques patented. Indicators of innovative activities, instead, enable us clearly to identify the use to which technology is put and therefore how countries differ in terms of industrial use of technology.

METHODOLOGY

Asking businesses to identify the typical and/or first user of their innovative outputs is a key feature of innovative activity surveys initiated in the mid-1970s (DeBresson and Townsend, 1978; DeBresson and Murray, 1984; DeBresson *et al.*, 1986). This feature was then included in the first survey performed in 1986 by a central statistical agency in Italy for 1981-85 and later in one taken in China for 1992. This makes it possible to compile innovative matrices in which each innovative activity is classified both under its supplier and its user industry. This corresponds to what Lundvall (1992) has called supplier-user interactions, interactions that are necessary in order to generate new ways of doing things and new techno-economic knowledge.

These innovative matrices are then compared with Leontief input-output tables for the specific country and period. The comparison yields robust statistical relationships, on the basis of which we were able to estimate an innovative activity matrix for France for 1990 and a more disaggregated one for China for 1992. Figures 1 to 5 present relative measures of innovative activities in terms of supplier-user inter-industry linkage for Italy, China and France.[3] For Italy and China, each industry supplying innovative outputs is made comparable (by dividing each cell by the total supplied by that industry), in order to see the share used by one industry or another. Differences among countries lie mainly in the importance of certain industries as users of innovative outputs, as reflected by the sharp peaks in Figures 1 to 5. For France, we used a slightly different means of comparison: in order to compare levels of innovative (Figure 4) and non-innovative activities (Figure 5), we measured the sales for both, which also enables us to compare levels of sales related to innovative activity among industries.

INNOVATIVE ACTIVITIES CLUSTER ONLY IN PARTS OF NATIONAL ECONOMIES

Innovative activities in all countries examined are located in only a small part of the national economies. Schumpeter suggested that this was so a long time ago (1912; 1939, pp. 99-100). His intuition has been abundantly confirmed in all studies to date. Although Schumpeter was talking about major innovations, we find that all innovative activity is clustered. This is so for Italy (Figure 1), China (Figures 2 and 3) and France (Figures 4 and 5). The case of France – with a conservative estimate – is particularly clear: sales related to innovative activities (Figure 4) cluster in very different industries than non-innovative sales (Figure 5).

THE DIFFERENT LOCATION OF INNOVATIVE CLUSTERS IN EACH COUNTRY

The main question that governments and investors want answered is: "Where in the economy are innovative activities concentrated?" Governments want to know

◆ Figure 1. ***The Italien innovative activity matrix, 1981–85***

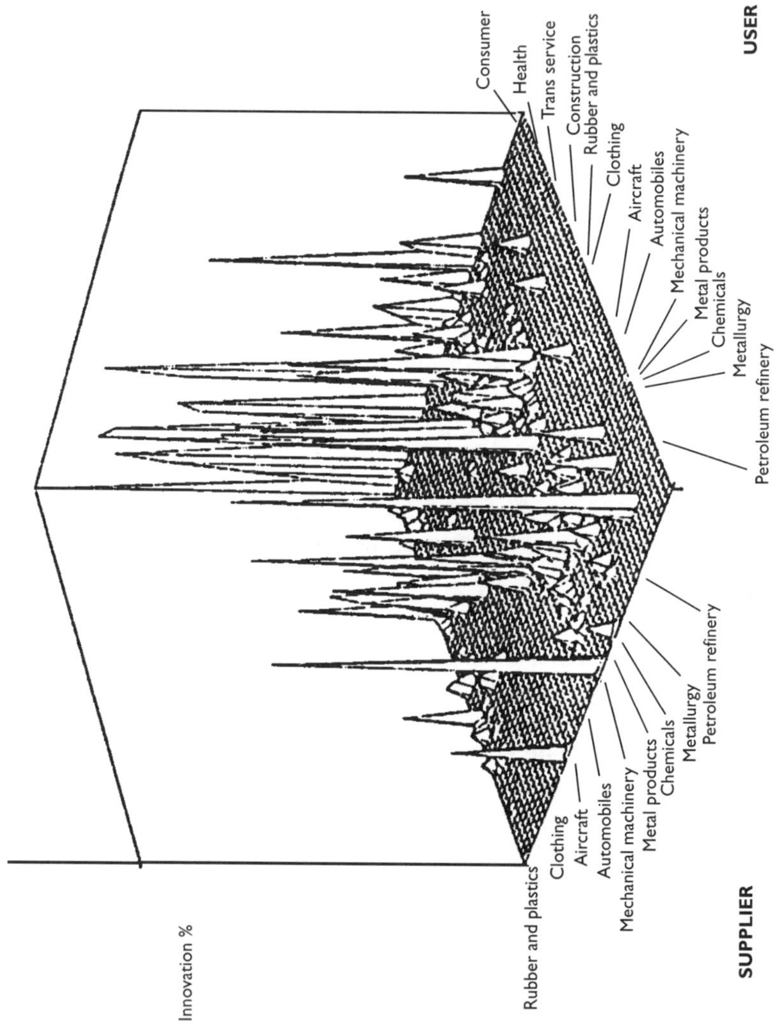

Source: Author.

◆ Figure 2. ***The Chinese innovation activity matrix, 1992***

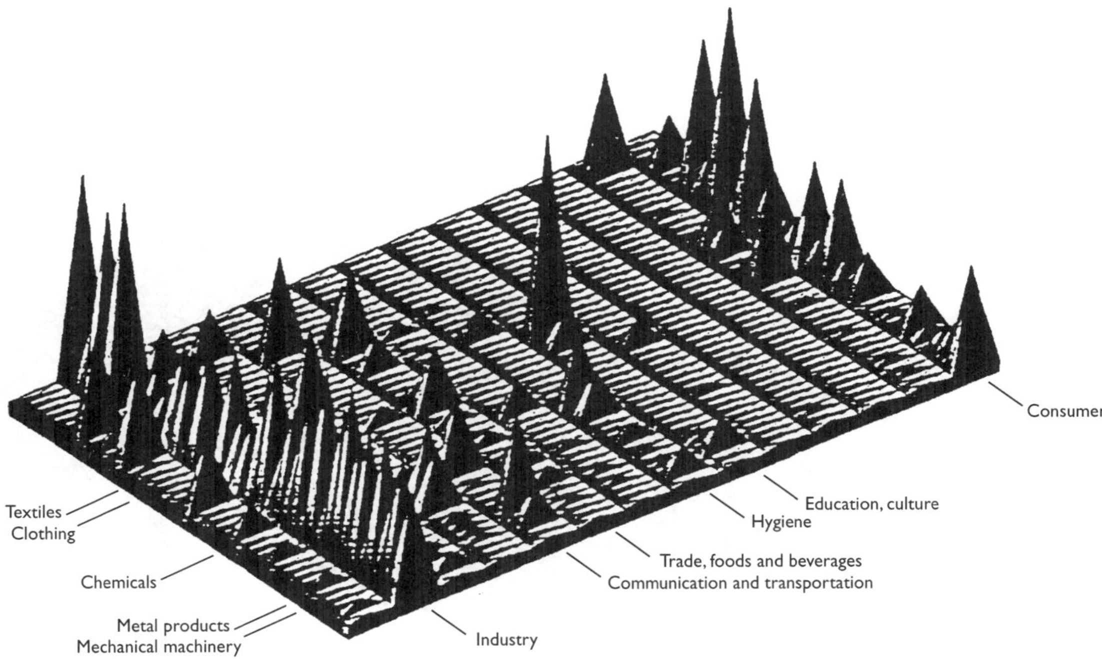

Source: Author.

◆ Figure 3. **The estimated inter-industrial matrix of innovative activity in China, 1992**

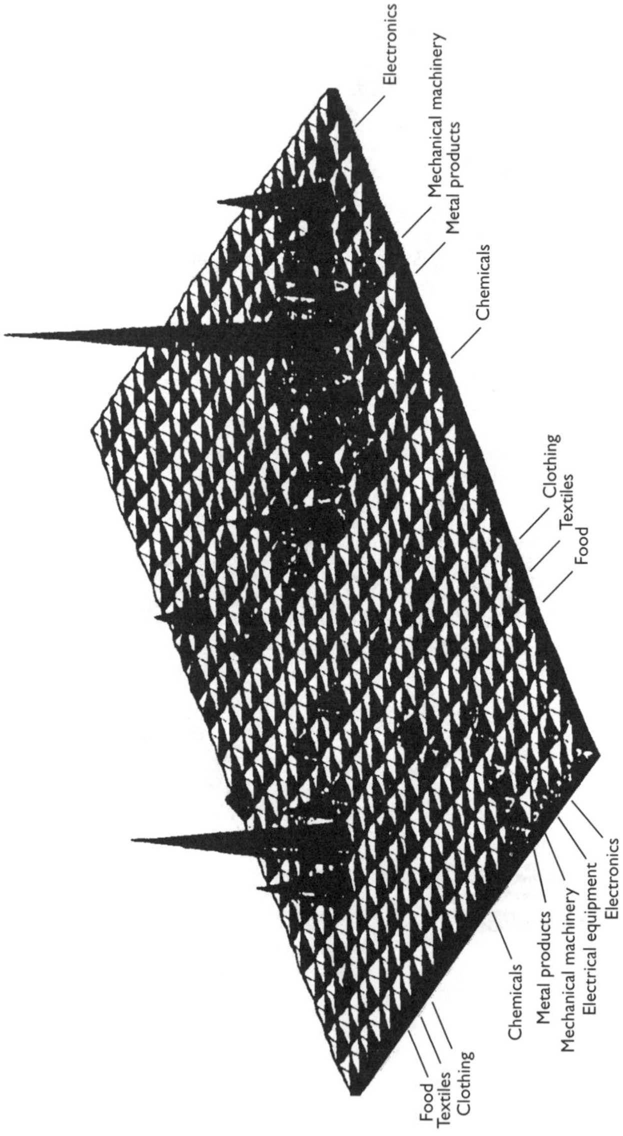

Source: Author.

◆ Figure 4. ***Innovative sales coefficients matrix in France***
(100 000 million francs)

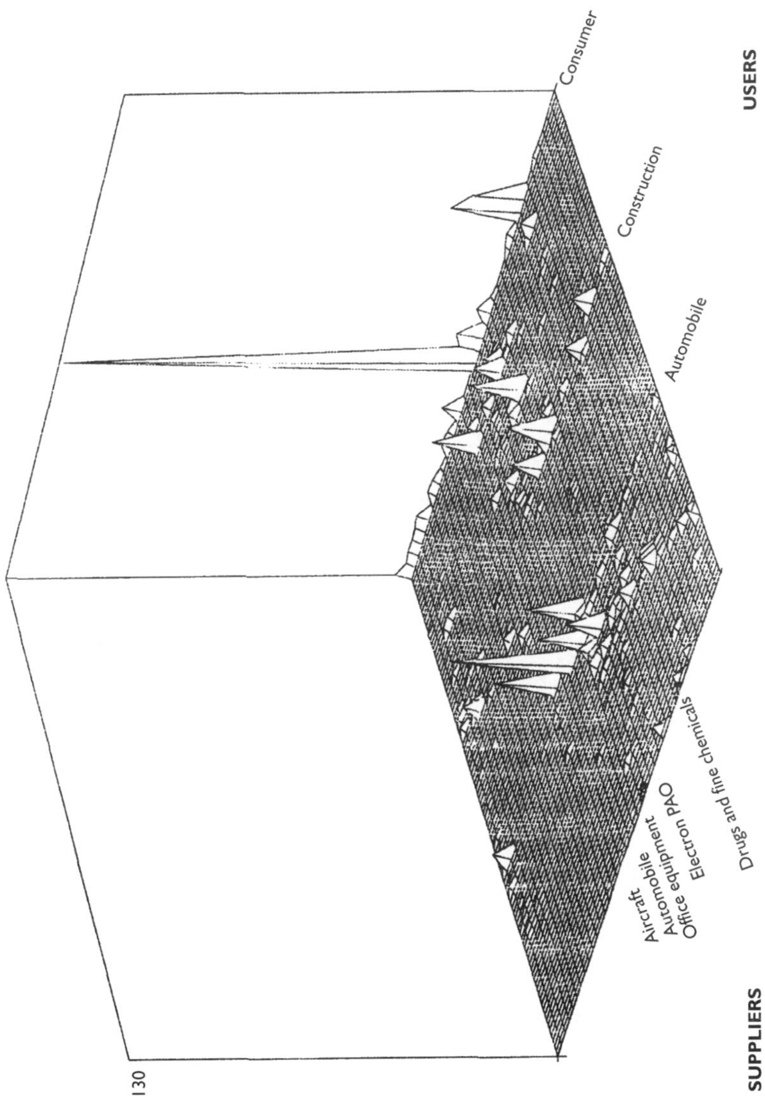

Source: Author.

◆ Figure 5. **Non innovative sales coefficients matrix in France**
(100 000 million francs)

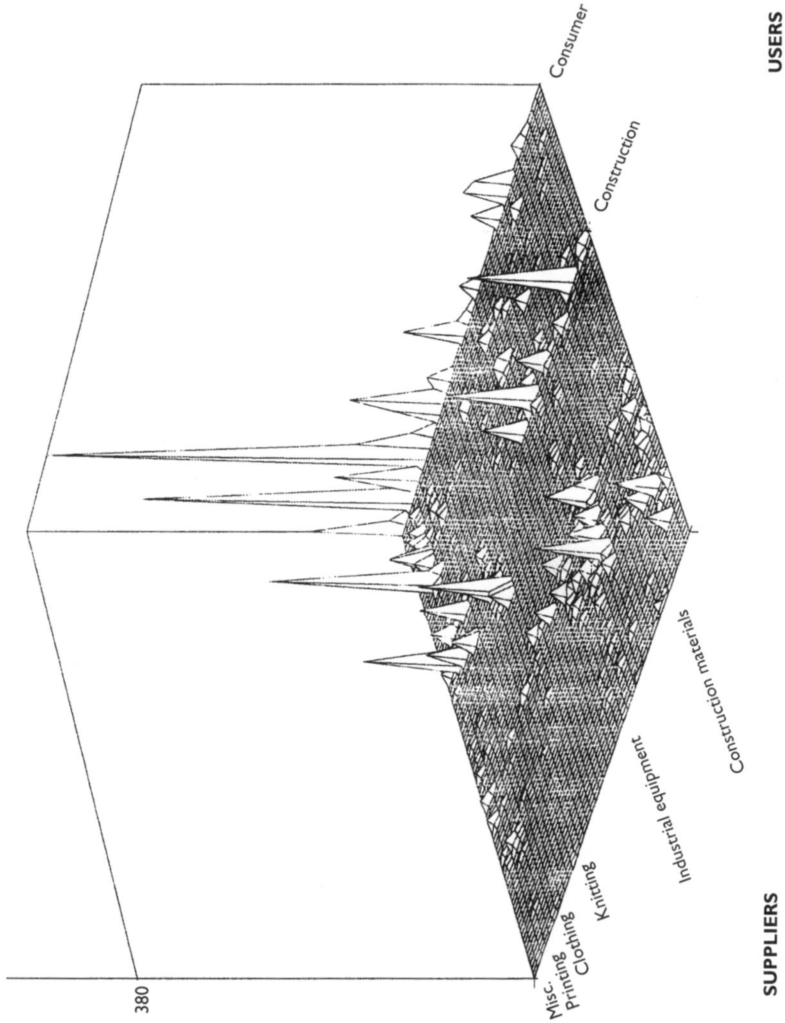

Source: Author.

the sources of dynamic specialisation. Investors want information to help them choose locations and competitive strategies. And while innovative activities cluster in all countries, they cluster in different places in different countries.

Italy

Figure 1 shows the matrix of innovative activity in Italy. The user industries are identified at bottom right; this shows that use of innovative activity during the period from 1981-85 was mainly concentrated in consumer goods and construction.[4] On the supplier side (bottom left), three industries seem to supply innovative activity to many user industries: mechanical machinery, chemicals and metallurgy.

China

The use patterns in China are different. On the final demand side (Figure 2) the most important user sectors in China are consumer goods, as in Italy, but also the health system, communications and transportation.

The more disaggregated manufacturing innovative matrix (Figure 3), clearly distinguishes three innovative activity clusters: the first revolves around machinery building, the second around the chemicals industry, and the third around the textile and food industries. On the supplier side, metal manufactured goods, as in Italy, and, to a lesser degree, chemical products, supply a large number of sectors with innovative products.

France

The patterns of use in France are different from those of Italy and China. Not only are innovative activities very centralised, they are located in economic areas very different from those where non-innovative activities prevail. In the case of France in 1990, the estimate is, as already noted, a conservative one, and innovative activities are probably even more clustered than in our estimate. In France, consumers are also main users, but innovative activity is more narrowly polarised around the automobile industry, fine chemicals, and pharmaceuticals (Figure 4). On the supply side, electrical equipment and appliances supply many industries with innovative outputs.

By comparison, enterprises that are not involved in any form of innovative activities are located in traditional industries: textiles, clothing and construction materials (Figure 5).

Table 1 summarises the different focal points for innovative activity in different countries; they are consistent with the known competitive and export specialisations of these nations. However, these clusters of innovative activities are not

Table 1. **Comparison of main centres of innovative activities in three countries**

Italy	China	France
Construction	Health services	Automobiles
Metal fabrication and mechanical machinery	Communications and transport	Fine chemicals
Chemicals	Metal products and mechanical machinery	Pharmaceuticals
	Chemicals	Electrical equipment and domestic appliances
	Textiles	

Source: Author.

strictly sector-specific. Although Table 1 identifies the clusters in terms of the user sector around which the innovative activity clusters, the graphs indicate, more accurately, that several related industries are involved in each innovative cluster. Industrial sectors cannot innovate alone. They need supplier industries for new components and user industries for new applications and new requirements. The most important implication is that competition through innovative activities concerns inter-related clusters of industries or networks of innovators – not isolated products, firms or industries.

A second remark is in order here. The same main supplying industries for innovations are found in two or three countries: mechanical machinery, metal fabrication, and chemicals. There is therefore not much difference among these countries for the supply of new technology. This is understandable: technological opportunities are related to scientific and technological discoveries that are international in scope; they do not depend only on the national economic setting. The differences are mainly on the user side or, more precisely, in the way in which suppliers and users combine their innovative efforts. The essential differences among countries reside in the way certain industries interact to innovate. Country specialisation rests on specific national systems of inter-industry innovative interactions. This confirms some of the intuitions and reinforces some of the hypotheses of the 1992 OECD report, *Technology and the Economy: The Main Relationships*; of recent studies of national systems of innovation (Lundvall, 1992; Nelson, 1993); and of the concept that nations acquire competitive advantages through internal linkages (Porter, 1990). It is likely, as we shall see in the next section, that inter-industry systems underlie these differences in the economic location of innovative clusters.

SIMILARITIES IN THE LOCATION OF INNOVATIVE CLUSTERS IN NATIONAL ECONOMIES

So far, we have stressed differences among countries. But are there some underlying similarities? How much of the difference in innovative clusters simply reflects a difference in economic structures? Does the relationship between economic structures and localisation of innovative activity show any recurrent patterns? If so, is it possible to foresee where in the economy innovative activity is most likely to emerge and cluster?

Figure 6 summarises our present answer to the above questions: the more backward and/or forward linkages an industry has, the more likely it is that it will have significant innovative activity. On the horizontal axis, we have plotted an industry's total number of direct backward linkages with domestic supplier industries and forward linkages with domestic user industries (calculated from the relevant national input-output matrices for corresponding or antecedent years). On the vertical axis, we have plotted measures of innovative activity. For Italy and China (left-hand vertical axis), we represent simple frequencies of innovative businesses in an industry.[5] For France, we present sales related to innovative activities.

An industry only has a high score for innovation when it has important domestic economic linkages (the upper right-hand corner of Figure 6): for example, the automobile industry for France, mechanical machinery for Italy and China. There are no high scores in the upper left-hand corner of the graph: very low domestic linkages are not correlated with highly innovative performances.

In general, the innovative performance of industries increases as domestic linkages increase, as the regression curves for each country show.[6] In all three cases, the beginning of an increase in domestic economic linkages is accompanied by only very slight increase of innovative activity; then, as the level of domestic linkages increase, the level of innovative activity augments quite dramatically.

This does not mean that all domestically well-integrated industries are highly innovative. The lower right hand corner of Figure 6 includes many industries, including the French paper and printing industries. This suggests that domestic economic linkages are not enough to induce innovative performance, but without them, highly innovative performance does not occur. Domestic economic linkages may be a necessary, but not sufficient, condition for innovative activity.

There are, of course, industries that do not conform very well to this picture and that have highly innovative performances with low levels of domestic integration (outliers above the regression curves in the middle or even on the left of Figure 6): the aircraft and drug industries are examples in a few countries (we will return to them below).

◆ Figure 6. ***Innovative activities and economic linkages***

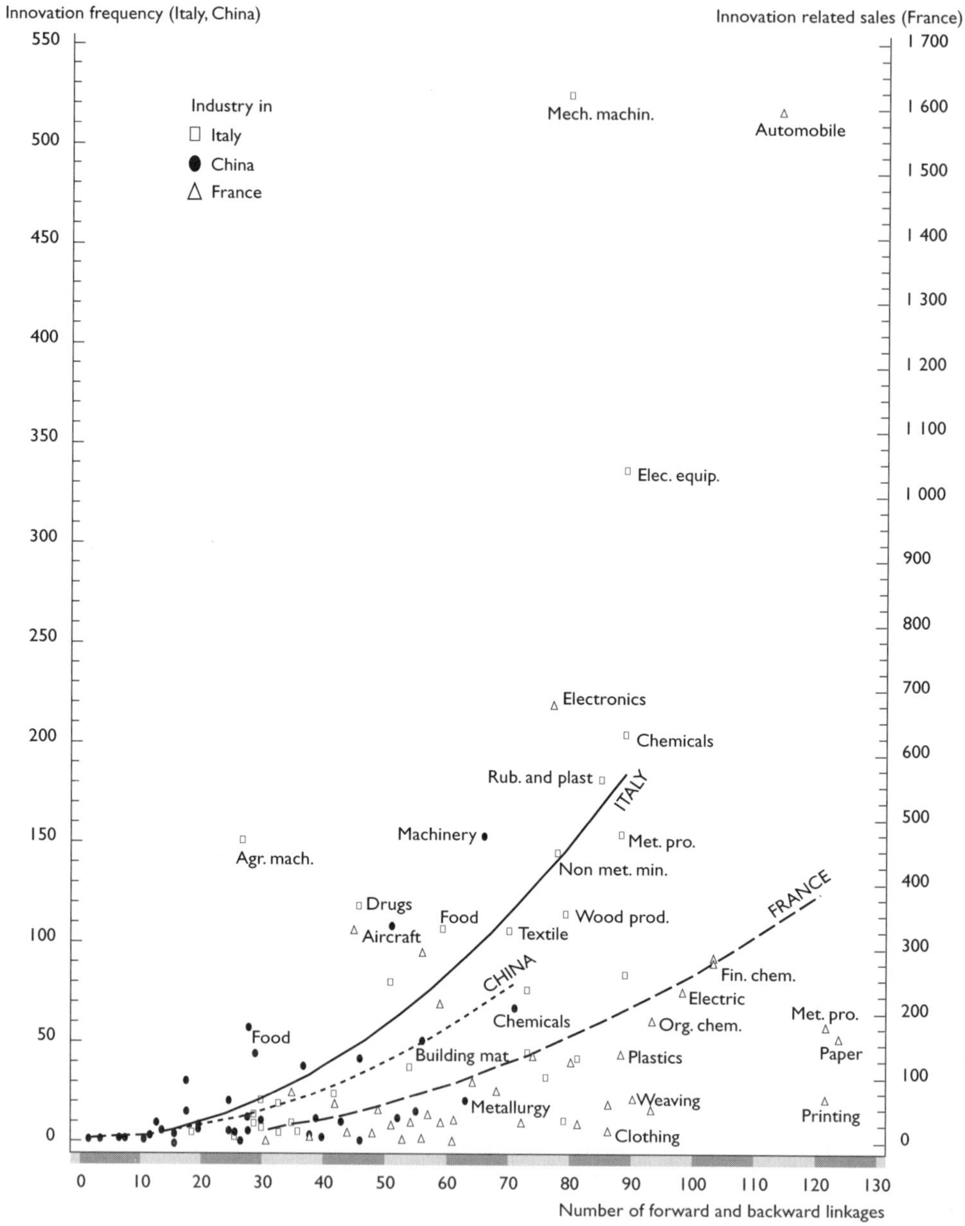

Source: Author.

A PLAUSIBLE INTERPRETATION: ECONOMIC LINKAGES MATTER FOR INNOVATIVE ACTIVITY

The most plausible explanation for the above relationship between economic linkages and innovative activity is a simple one: the more varied the enterprise's information network is, the more likely it is to combine production factors in a new way for new uses. Innovation, as Schumpeter stressed, requires first and foremost new combinations. Obviously, the number of possible combinations is a function of the variety of elements to be combined. If the entrepreneur is in contact with a variety of potential input suppliers, he has more opportunities to innovate. Conversely, if he has a a variety of potential market outlets for his products, he has greater possibilities for innovative variation.

This economic rationale explains why backward and forward linkages are favourable to innovative activity. These linkages might include foreign linkages through imports, technology transfer from abroad, foreign direct investment, exports of goods, or domestic technical know-how.[7] It does not explain why domestic linkages are important. So far, we have only looked at domestic linkages. The recurrent outliers in Figure 6 – aircraft and drugs – may be more influenced by foreign linkages: we know that the aircraft industry supplies itself with critical components world-wide and that the drug industry taps medical research abroad.

Nonetheless, it is possible that in industries that are not science-related, where knowledge is less codified, transaction costs – and innovative interaction between firms is transaction-intensive – may be lower for innovative partners when they belong to the same national culture and share the same language and customs, as well as the same legislative and regulatory framework. This may help give importance to domestic economic linkages as a necessary condition for innovative combinations.

The above findings, however, beg the following question: what are the types of linkages that encourage innovative activity, and at what level (national, international, regional) are they significant? The question has yet to be answered.

For technological and economic development policy, the above patterns have important implications. For a firm's strategic decision making (investment location or design of supplier network), these findings also have important implications. Competition through innovation does not, according to this view, occur between isolated enterprises or industries as the atomistic vision of standard economics – be it neo-classical or Schumpeterian – suggests. It occurs between networks of enterprises, which innovate through interaction and create new technological know-how together. They co-operate to compete for resources and markets against other networks. Some of these networks are anchored in national economic spaces, even when they compete world-wide. For innovative enterprises, the immediate economic environment may affect how well they interact innovatively with others and constitute the crucial networks.

NOTES

1. This work was presented in preliminary form at the OECD's Group of National Experts on Science and Technology Indicators (NESTI) Workshop on Innovation, Patents and Technological Strategies, held in Paris in December 1994. It highlights findings of a forthcoming book, *Economic Interdependence and Innovative Activity: An Input-output Analysis*, Edward Elgar Publishers. We would like to thank Giorgio Sirilli, Philippe Kaminski and Xu Shiqing – co-authors of the book – for their contributions, as well as an anonymous referee and François Pham for their comments to this summary.
2. Not all surveys have yet explicitly identified the industrial user of innovative outputs. This dimension has, however, revealed itself as one of the most telling elements of innovative activity surveys. This chapter demonstrates the usefulness of identifying the users as well as the suppliers of innovative outputs.
3. The actual matrices are too unwieldy and cumbersome to be presented here.
4. In all the countries, each industry uses in its processes a significant proportion of its own innovative outputs: this is reflected on the diagonal running from lower centre to the back centre left.
5. Other measures – percentage of sales, degree of novelty, costs – present the same distributions.
6. Although we do not supply the many statistical indicators, we only present results when they are all statistically significant and not sensitive to different measures of innovative activity.
7. In the case of Italy, where our data is finer, we found some evidence of a correlation between imports of goods and foreign technical know-how and innovative performance.

REFERENCES

DeBRESSON, C. and J. TOWNSEND (1978), "Note on the Interindustrial Flow of Technology in Postwar Britain", *Research Policy*, 7(1), pp. 48-60.

DeBRESSON, C. and B. MURRAY (1984), *Innovation in Canada: A Report to the Science Council of Canada,* 2 vols., New Westminster, CRUST reprint.

DeBRESSON, C., B. MURRAY and L. BRODEUR (1986), *L'Innovation au Québec*, Les Publications du Québec, Quebec.

JEWKES, J., D. SAWERS and R. STILLERMAN (1958), *The Sources of Invention,* MacMillan, New York, 2nd rev. edn. (1969), Norton, New York.

LUNDVALL, B-Å. (ed.) (1992), *National Systems of Innovation: Towards a Theory of Innovation and Interactive Learning*, Pinter Publishers, London.

NELSON, R.R. (ed.) (1993), *National Innovation Systems: A Comparative Study,* Oxford University Press, Oxford.

OECD (1992), *Technology and the Economy: The Key Relationships*, Paris.

PORTER, M.E. (1990), *The Comparative Advantage of Nations*, Free Press, New York.

SCHUMPETER, J.A. (1912) "Theorie des wirtschalisches Entwicklung Berlin", in J.A. Schumpeter (1939), *Business Cycles: A Theoretical, Historical and Statistical Analysis of the Capitalist Process*, 2 vols, McGraw-Hill, New York.

9

RELATIONSHIPS BETWEEN MULTINATIONAL FIRMS' TECHNOLOGY STRATEGIES AND NATIONAL INNOVATION SYSTEMS: A MODEL AND AN EMPIRICAL ANALYSIS[1]

by

Rémi Barré

Observatoire des Sciences et des Techniques, France

INTRODUCTION

Over the last twelve years, a reappraisal of Schumpeterian theories has led to a new body of knowledge and a better understanding of the innovation process. The central concepts of externalities, learning processes, non-market and non-hierarchical interactions among various agents, skills accumulation, trajectories, path-dependency and irreversibilities have been incorporated into a coherent framework involving evolutionary theories of technical change at the microeconomic level and endogenous growth theories at the macroeconomic level. Links with contracts theory, on the one hand, and with regulationist theory, on the other, have also recently opened further and promising avenues for investigation.

Much work has recently been devoted to two phenomena that are at the centre of many policy debates: the national system of innovation (NSI) and the multinational enterprise (MNE). Remarkably, these two lines of investigation have developed quite independently, and their findings have seldom been related and placed in an overall framework. On the one hand, we understand better why and how geography, institutions, culture and history shape the specific dynamics of innovation in national systems; on the other, the MNEs' rationale for world-wide innovation strategies have been thoroughly discussed and documented.

Using a quantitative approach, this chapter seeks to relate these two bodies of knowledge in order to cast new light on both. Its goal is to show that the technological strategies of MNEs result in the establishment of international innovative networks (IINs), which interact in a significant manner with the NSIs of both home and host countries. The hypothesis is that the different strategies used by MNEs with respect to the volume and technological orientation of the innovation networks they build with their subsidiaries abroad reveal the characteristics of both home and host NSIs.

Patel and Pavitt (1991) have shown the importance of home-country NSIs for the technological performance of large firms; we propose to complete the model to show how MNE innovation strategies feed back into NSIs.

We first discuss the rise of the IIN among and within MNEs. Next, we present a conceptual model of the interaction between IINs and NSIs of both home and host countries. Finally, we assess the situation of 11 countries plus the European Union (EU) with respect to their MNE strategies.

RATIONALE FOR AND SCOPE OF INTERNATIONAL INNOVATION NETWORKS

The internationalisation of the economy and the MNE[2]

Internationalisation – or "globalisation"[3] – is the interaction of two phenomena: the mobilisation of resources at world scale and the international dimension of economic competition. It has been a characteristic of the evolution of the world economy over the last decades, but during the 1980s, the deregulation of financial markets, the liberalisation of trade, and the development of communications technologies have accelerated the trend.

MNEs are the major actors in the internationalisation of the world economy, since they control and co-ordinate value-added activities across national boundaries through foreign direct investment (FDI) international trade of intermediate goods, and inter-firm agreements (Sachwald, 1993; Maarten de Viet, 1993).

Much work has been done on analysis of the factors explaining the multinationalisation of the firm.[4] The basic idea is to analyse how the firm – considered as a set of capacities – increases its competitive advantages through multinationalisation. Each firm's strategy will depend on its competitive position, on the characteristics of the competition in its sector and of its home country. The specific forms of organisation will finally be determined by the trade-off between the benefits of integration and the costs of co-ordination and control.

In practice, the relationships between the entities that constitute the MNE are defined by a combination of financial, contractual and organisational characteristics. Most involve firms with financial links (between a central firm and its subsidiaries and/or among subsidiaries), but purely contractual linkages – co-operation or agreements – became more important during the 1980s. Like all other functions, technology and innovation are affected by multinationalisation and have sometines stimulated it (Howells and Wood, 1993).

IINs and MNE strategies

An MNE internationalises its innovation activities in three principal ways:

- access to and use of basic research results at world scale through: *i)* scientific journals; *ii)* the "invisible colleges" to which industry researchers may belong, as long as they comply with the norms of disclosure of academic research; and *iii)* contractual relationships between the firm and academic institutions;
- technological co-operation agreements between firms at international level;
- creation of IINs within MNEs, *i.e.* the body of interactive innovation activities taking place between a firm (at home) and its subsidiaries abroad (in host countries).

We will focus here on the third of these.

Because the technology and innovation functions fundamentally depend on generating and combining knowledge, their delocalisation at several sites necessarily implies establishing innovation networks within the MNE; in what follows, IINs are the internationally dispersed units which, together, are charged with implementing MNEs' technology and innovation functions.[5] Such IIN result from FDI related to technological activities.

Finally, the diversity of strategies, management principles, and organisation in an MNE's innovation function can be grouped in terms of three contrasting types, each of which is also characterised by one kind of innovation network.[6]

Innovation strategy A: home-based R&D and local adaptation networks

The firm of origin centralises most of the R&D, and the role of the subsidiaries abroad is limited to adapting the products to local markets. IINs are quite weak, and functional integration quite limited.

Innovation strategy B: networks based in the host country

R&D and innovation are centralised but not in the firm of origin: a subsidiary, because of its size and capabilities, takes responsibility for the innovation function, for certain product lines or technological areas. IINs are quite important, and functional integration quite extensive.

Innovation strategy C: division of labour networks

The various capabilities that participate in the innovation process are located where they are at their best: the various sites are complementary and there is division of labour. IINs are important, and functional integration very extensive.

THE RELATIONSHIPS BETWEEN MNE STRATEGIES AND THE HOME/HOST COUNTRY'S NSIs

The influence of home/host countries' NSIs on MNE strategies

The model can be specified in terms of the relationships between the strategy of MNEs and the characteristics of home and host countries (see Table 1). When the home country is both industrially and technologically strong, MNEs need not look abroad for technological capabilities and can limit themselves to strategy A; when dealing with technologically strong but industrially weak countries, they will be interested in buying firms with good technological assets (strategy B). MNEs in countries with strong technology but weak industry will behave similarly. Countries with weak technology (or a small NSI) will tend to develop strategies B and C towards strong technology countries, if they can afford to.

Table 1. **MNE strategies as a function of the characteristics of home and host countries' NSIs**

Home country of MNEs	Host country of MNEs' R&D subsidiaries		
	Strong industry and strong technology	Weak industry and strong technology	Weak technology and/or small NSI
Strong industry and strong technology	Strategy A	Strategy A or B	Strategy A
Weak industry and strong technology	Strategy A	Strategy A or B	Strategy A
Weak technology and/or small NSI	Strategy A, B or C	Strategy A, B or C	Strategy A

Source: Author.

The influence of MNEs' innovation strategies on the volume and technological orientation of their IINs

The relationships between MNE strategy and IIN size and technological orientation can be summarised in the following way (see Table 2):

– local adaptation *(strategy A)*: small volume IINs, in the same technological area as the home-based R&D facilities, since the local facilities in the host country develop and adapt the results of technology developed in the home country;

Table 2. **Relationship between MNEs' innovation strategies and the volume and orientation of their IINs**

Characteristics of IINs	MNEs' strategies		
	Strategy A Local adaptation	Strategy B Host-country centralisation	Strategy C Division of labour
1. Volume Volume of NOUT(i)	Small	Large	Small or medium
2. Technological orientation Comparison of technological specialisation pattern between:			
– home country and NINW(i)	No link	No link (similar if large-scale B)	Similar
– home country and NOUT(i)	Similar	Similar	Opposite (similar if country has small NSI)

NOUT(i) Technological activities taking place outside of the home country (i) of the MNE, generated by outwards investments by the country's MNEs. NOUT(i) refers to the parts of the IINs of the MNEs of the country (i) which are located outside the country (i).
NINW(i) Technological activities resulting from inward investments of foreign MNEs in the country (i) considered as host country; NINW(i) refers to the parts of the IINs of the MNEs of foreign countries that are located in the host country (i).
Source: Author.

- host-country centralisation *(strategy B)*: significant IINs, in the same technological area as the home-based R&D facilities, since host-country facilities tend to complement home-based facilities;
- division of labour *(strategy C)*: medium or modest IINs, in technological areas different from those of excellence of the home country, since the aim is to access capabilities not found in the home country; in the case of a small NSI, the technological areas of an IIN may be the same as areas of excellence at home: what is sought abroad is extension of existing, but too small, local capabilities.

The feed-back loop: the impact of IIN-MNE on the countries involved

The volume and type of an IIN depend on the NSI of both home and host country. The point made here is that: *i)* NSIs are also affected and modified by IINs; and *ii)* these interactions differ, depending on the characteristics of the NSIs.

We will not return here to the notion of NSI, but we recall, in the words of Patel and Pavitt (1991), that there are country-specific conditions – the financial, educational and management systems, and degree of competitive rivalry – that generate spillovers. They constitute the basis of cumulative and irreversible processes of technological, organisational and market learning and development of the know-ledge and skills of people and institutions. These conditions and processes shape national competitive advantages and have a major influence on the technological activities and directions (strategies) of firms, and they are reflected in international patterns of trade and production.

We now turn to the feed-back effects of the strategies of the MNE on the various NSIs in which they develop their IINs as a result of these strategies. The question of the significance of the IIN for the dynamics of NSIs has usually been addressed in qualitative terms[7] (Cantwell, 1989; Harris, 1991; Dunning, 1994); we outline below the basic features of such effects.

Specialisation effects (strategies B and C)

MNEs tend to set innovation activities –- that is, to develop their IINs – in sites with a technological specialisation and thus a technological advantage in their sector of activity. In this way, the firm will be able to mobilise local capabilities, including the externalities that are specific to the site (Cantwell, 1989 and 1991). In other words, "non resident-owned research tends to be drawn to those sectors in which a recipient country has a comparative patenting advantage" (Cantwell and Hodson, 1990). The MNEs whose home countries are too narrowly specialised – having, for example, small NSIs – will tend to extend their innovation activities abroad (Patel and Pavitt, 1991; Cantwell and Hodson, 1990). Thus, using innovation strategies B and C, they will primarily create linkages with centres of excellence in host countries.

Dynamics of vicious and virtuous circles: convergence/divergence effects

Beyond this general framework, it is important to see that the impacts of the IIN on the host country will be to some extent country- and industry-specific, and that there are also indirect effects, in the form of spillovers. For direct effects, the question is the degree to which IIN links to foreign firms complement or replace locally based networks and innovation activities. Indirect effects – the spillover effects on suppliers and customers, local competitors, R&D personnel – are wide-spread; they may speed the process of growth and restructuring (Dunning, 1994) and raise the overall level of capabilities, or they may focus and shape those capabilities in ways that are incompatible with the characteristics of local (host country) industry. In fact, it appears that contrasting effects appear, depending on the strength of the national system of innovation :

- If local industry is strong, there are chances that an IIN from abroad, in its direct and indirect effects, will generate a virtuous circle: stimulation, competition, beneficial externalities, all resulting in increased attractiveness.
- If local industry and technology are weak, a foreign-based IIN may well tap into unique local R&D resources, focus on problems of no relevance to the host economy, and out-compete local industry. The effects are those of a disguised "brain drain". But foreign IIN generated by FDI, especially in the case of strategy A (local adaptation), can generate beneficial spillovers, for example by upgrading the standards of local suppliers; countries like Korea in the 1970s, Taiwan and Singapore in the 1980s and China, Thailand and the Philippines in the 1990s are following this route to upgrade their NSIs.

The problem is one of asymmetries between countries (Granstrand *et al.*, 1993), with one country's vicious circle becoming another's virtuous circle, so that weaker countries either drift away or catch up. The differentiation among countries and/or regions in terms of technological capabilities and specialisation may tend to increase, as technological capabilities tend to cluster in certain sites. This may result in the strengthening of existing poles of excellence (Maarten de Viet, 1993), and, symmetrically, the weakening of weaker countries or regions.

Home country NSIs can be oriented towards certain trajectories, which may, or may not, correspond to their own goals. This can have negative effects if the result is to leave aside important technologies or technologies that are key to activities that have a growing market.

The effects on the presence or absence of countries in strategic industries

Some industrial sectors are based on technologies with rich opportunities and high prospects for future growth and profits. They are characterised by high entry barriers, even in the long term, and important scale economies and sunk costs in R&D. In such imperfectly competitive markets, there exist oligopoly rents, even in the long term. Such sectors are often labelled "strategic" for economic and overall security reasons.

The issue is to know under what conditions at least one firm in a country belongs to such oligopolies. It would appear that MNE strategies may play a decisive role in determining whether countries enter – or fail to enter – the "club" of those that master the strategic activities of future economic competition.[8] The various interactions among national innovation systems through the activities of the IINs of of MNEs are summarised in Table 3.

Table 3. **Interaction between countries' innovation systems through the IIN activities of MNEs**

Home country of MNE	Host country of MNE subsidiaries		
	Strong industry, strong technology	Weak industry, strong technology	Weak technology and/or small NSI
Strong industry and strong technology	Virtuous circle for both countries	Sale of technological assets by host country	Potentially vicious circle for host country
Weak industry and strong technology		Sale of technological assets by host country	Potentially vicious circle, unless carefully designed policy
Weak technology and/or small NSI	Catch up by home country; but cost may be high		

Source: Author.

A conceptual model of interaction between IINs and NSIs

The elements of knowledge identified in the above review are:

- NSIs have a major influence on MNE strategy.
- R&D abroad (*i.e.* the foreign IIN of an MNE or outward IIN), a major component of MNE strategy, can differ widely in volume even among MNEs of countries of similar technological strength. This is due both to the inter-firm variability of strategy (linked to their micro-scale specificities) and to the variation in the size of the NSI (MNEs with small NSIs will tend, other things being equal, to need to develop more IIN abroad).
- When they choose to extend their IIN abroad, MNEs tend to choose countries strong in the particular technology they are interested in. Nevertheless, part of the IIN consists in facilities for adapting technology; these are located in the market countries, not in those of technological strength.
- There are feed-back mechanisms which consist in the influence of the foreign (inward) IIN on the host NSI, the nature of which depends both on the kind of foreign IIN located in the country and on the characteristics of the receiving (host) NSI itself.

Finally, MNE strategies regarding IINs depend on the NSIs of both home and host country, and have an impact on both; Hence, the model presented in Figure 1.

◆ Figure 1. ***A representation of the interactions between IINs and NSIs***

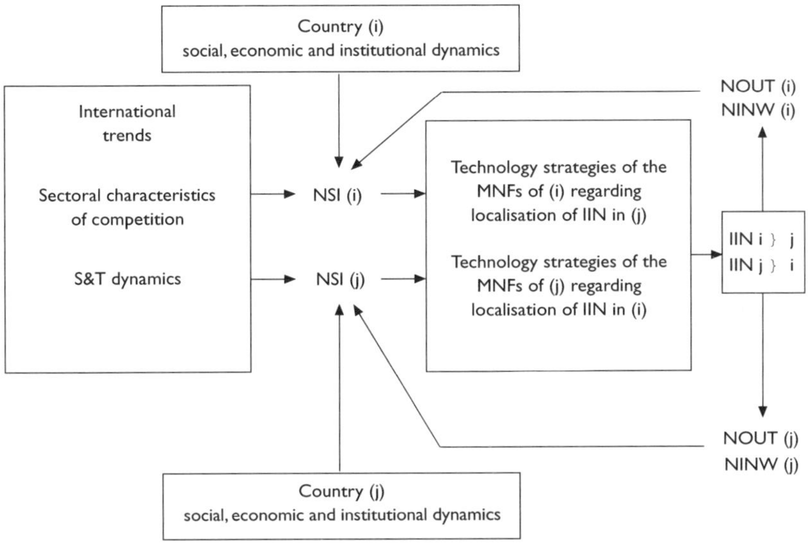

Notations: NSI (i): national system of innovation of country (i);
MNF (i): multinational firms from country (i);
IIN i } j: international innovation networks established between the home country (i) of the MNF (i) and their subsidiaries located in country (j);
NOUT (i): technological activities of all subsidiaries of MNF (i) located abroad
= summation of IIN i } j for all j's;
NINW (i): technological activities of all foreign subsidiaries located in country (i)
= summation of IIN j } i for all j's.

Source: Author.

The characteristics of the framework are that MNE strategies with respect to the innovation function depend on both home and host countries and that the resulting IIN will affect both home and host countries. Beyond this very general framework, the nature of some of the interactions can be specified. The suggests that characterising both the inward and outward IIN by size and technological orientation can:

– reveal an MNE's prevailing opinion about the country, and about therefore the status of the country's NSI;
– help analyse the possible implications of internationalisation of innovation for that country.

A QUANTITATIVE APPRAISAL OF THE INTERACTIONS BETWEEN IINs AND NSIs FOR 11 COUNTRIES PLUS THE EUROPEAN UNION

Methodologies used in previous work

In recent years, a number of studies, usually consisting of an enquiry addressed to a few or a few dozen firms, have been carried out.[9] These studies have all the values and limitations of case studies. The issue here is a reliable appraisal of *all* IINs.[10] The basic idea behind the few analyses done with this objective is to consolidate (aggregate) all the firms in the world that are part of an industrial group, *i.e.* that are subsidiaries of the same holding company. The hypothesis is that, beyond a certain threshold of financial participation, a firm, along with its innovation activities, becomes part of a management network composed of the various subsidiaries and of the holding company. The analysis then consists in measuring and consolidating classical parameters such as R&D expenditure and patenting.

For consolidated R&D expenditure, the problems are, first, the well-known limitations of using R&D expenditure as a proxy for innovative activities, and, second, the non-accessibility of data at firm level, especially when the aim is to cover all countries.[11] This is why the only published results concern a single country – the one for which the analyst had access to the data (for example, Madeuf *et al.*, 1993, for France; and National Science Board, 1994, for the United States).

For consolidated patent indicators, there are, of course, the intrinsic limitations of the patent indicator, but here it is possible to carry out systematic and exhaustive analysis at world level. Such complete (or near complete) analysis was pioneered by Patel and Pavitt in the late 1980s on US patenting. They have since launched several studies on MNEs and their relationships with NSIs (for example, Patel and Pavitt, 1991). Results have since been obtained on European patents[12] (OST, 1993).

The networks postulated by the subsidiary consolidation method are both too wide and too narrow. They are too wide, since some subsidiaries may have total technological independence (for example, when they are in fields unrelated to the holding company). They are too narrow, since innovation networks exist among firms without any financial linkages (in the case of technology agreements, for example). Among the means available to overcome these difficulties, there are the direct use of an inter-firm technological agreement database (Hagedoorn and Shakenrad, 1992),[13] the use of a co-patenting indicator,[14] or the analysis of the technological proximities and complementarities among firms in the same industrial group.

An appraisal of IINs through the European patenting indicator

We present here four indicators for appraising the IINs of 11 countries for all technologies.[15] They have been updated and reworked from those already published (OST, 1993).

Principles of measurement of technological activities

In order to appraise the volume and orientation of IINs with respect to NSIs, we make two basic assumptions:

- patenting by firms is a proxy for their technological activities, both in volume and orientation;
- affiliated firms of MNEs are a proxy for the latters' "networks".

Thus, patenting activities of the foreign affiliates of a country's MNE measure its outward technological investment and activities (NOUT). Reciprocally, patenting by the national affiliates of a country's foreign MNE measures inward technological investments and activities (NINW). The various parameters can then be defined and measured through proxies, as presented in Table 4.

The database[16]

The database used was built from European patent data: for all firms with at least six patents in 1989-90 (publication years) – 2 021 firms – information was obtained on financial linkages. 384 of these firms are single firms, the remaining 1 637 firms are related to one of 703 multinational groups. Each of the latter has a core firm, whose nationality is the nationality of the group. Finally, each patent applied for by one of the 2 021 firms can be related to: *i)* the nationality of it(s) inventor(s); *ii)* the nationality of the institution(s) to which it has been granted; and *iii)* the nationality of the group that "controls" the patent (the group to which the applicant institution belongs).

The total number of patents published for the years 1991, 1992 and 1993 is 171 274. For 98 794 of them (58 per cent), the applicant is one of the 2 021 firms considered here; 96 888 of them originated in one of the 11 countries under consideration. Our analysis covers 57 per cent of all European patents published during the period 1991-93. Patents not considered are those:

- for which the applicant is not an institution;
- for which the applicant is a small institution in terms of technology production (less than six patents published for the period 1989-91);
- for which the inventor or applicant does not belong to one of the 11 countries considered.

Table 4. **Parameters of the model: code, name, definition and proxy used for measurement**

Code	Name	Definition	Proxy
NOUT(i)	Outward-oriented IINs of country (i)	Technological activities of the MNE's IIN of country (i) located outside country (i)	Number of patents from subsidiaries of MNEs of (i) which are invented outside country (i)
NINW(i)	Inward-oriented (towards i) IINs	Technological activities of IIN of MNEs of foreign countries located in country (i)	Number of patents invented in country (i) for firms that are affiliates of foreign MNEs
NDOM(i)	Domestic technology for national firms	Domestically produced technology for national firms of country (i)	Number of patents invented in country (i) for firms of country (i)
NX(i) = NDOM(i) + NINW(i)	Technology created by country (i)	Technology produced in country (i) by subsidiaries of foreign firms and by domestic technological activities of national firms	Number of patents by inventors located in country (i)
NY(i) = NDOM(i) + NOUT(i)	Technology "controlled" by the firms of country (i)	Technology produced abroad and "controlled" by firms of country (i) and by domestic technological activities of national firms	Number of patents by inventors located abroad, in affiliates of firms of country (i) plus patents invented in country (i) by national firms

Source: Author.

Results regarding the volume of IINs linked to MNEs (IIN-MNE)

More than 20 per cent of Canadian and Belgian industrial laboratories patenting in European are controlled by foreign firms; the same is true for more than 15 per cent of UK labs (Table 5). In Japan, on the other hand, only 1.7 per cent of European patents are controlled by non-Japanese firms. The corresponding figure is 4.8 per cent for the United States and and 4.9 per cent for the EU. These results can be compared with the results of studies on the share of R&D and US patenting performed in subsidiaries of foreign firms (Table 6). The results of the three different analyses generally converge.

Table 5. **INW-IIN indices**

Percentage of patent by resident inventors for foreign firms

	F	D	UK	I	ND	B	SW	CH	CAN	US	J	EU
INW-IIN	7.4	5.3	15.5	6.2	11.5	29.2	8.0	11.6	23.9	4.8	1.7	4.9

Source: Author.

Table 6. **Comparison with results from other quantitative studies**

%	F	D	UK	I	ND	B	SW	CH	CAN	US	J	EU
INW-IIN	7.4	5.3	15.5	6.2	11.5	29.2	8.0	11.6	23.9	4.8	1.7	4.9
% R&D in foreign subsidiaries [1]	12		17				14		44	7	1	
% US patents in foreign subsidiaries [2]	11.8	11.5	22.3	11.2	9.5	45.7	5.4	12.5	28.1	4.2	1.2	7.4

Sources:

1. OECD (1993), p. 53, EAS database (year 1987).
2. OECD (1991), from Patel and Pavitt, *US Patenting from Inside Country by Foreign Firms* (as percentage of country's total US patenting).

••• **Outward IIN index** OUT-IIN (i) = NOUT/NDOM

from 0 (no external IIN) to 1 (all patents controlled by domestic MNE are invented in their foreign subsidiaries)

As Table 7 shows, IINs controlled by Dutch and Swiss firms have more than 25 per cent of their technological activities located abroad. For Japan, only 0.8 per cent of the technological potential is located abroad, as far as European patenting is concerned; the United States controls almost twice as many labs in foreign countries as does the EU.

Table 7. **Outward IIN indices**

Percentage of patents by foreign inventors among total patents controlled by domestic firms

	F	D	UK	I	ND	B	SW	CH	CAN	US	J	EU
IIN-B	4.6	4.8	9.8	2.0	28.7	9.6	7.7	26.5	10.0	7.1	0.8	3.8

Source: Author.

••• **IIN balance index** IIN-B (i) = NINW/NOUT

higher (lower) than 1: inventors of (i) invent more (less) for foreign firms than foreign inventors invent for domestic firms; IIN-B > 1 means that country (i) is a net exporter of inventions

This indicator compares the volume of foreign networks in the country to the volume of networks controlled by the country but located abroad. As Table 8 shows, only three countries – the Netherlands, Switzerland and the United States – have a balance ratio below unity, meaning that the IINs that they control abroad is bigger that the IINs that foreigners control in their country. These three countries are net importers of technology. On the other side, Italy, Canada and Belgium are the largest relative exporters of technology, with a balance of above 2. The EU and Japan are exporters of technology, while the United States is an importer.

Table 8. **IIN balance indices**

Ratio of patents by domestic inventors for foreign firms to patents by foreign inventors for domestic firms

	F	D	UK	I	ND	B	SW	CH	CAN	US	J	EU
IIN-B	1.6	1.1	1.6	3.1	0.4	3.0	1.0	0.4	2.4	0.7	2.1	1.3

Source: Author.

The following main conclusions can be drawn from these indicators:

- for small highly industrialised countries plus the United Kingdom, the IIN phenomenon is substantial, involving between 10 and 30 per cent of their technological activities;
- for the bigger countries, the IIN phenomenon accounts for 5 to 8 per cent of technological activities, except in Japan, where it is negligible;
- countries have highly specific and contrasted patterns for IINs.

Relation of the technological orientation of IINs to characteristics of NSIs

Statistical treatment performed

We use the data from tables NOUT and NINW per country (i) per technological area (k).[17] If N(i,k) is the measurement of a parameter N characterising country I on the technological area k,[18] let us call n(i,k) the following index:

$$n(i,k) = [N(i,k) / N(i,.)] / [N(.,k) / N(.,.)]$$

n (i,k) is the vector of "specialisation indices" of country (i) related to parameter N; it is sometimes also called "revealed comparative advantage".

Let us call nDOM, nINW, nOUT the indices computed from NDOM, NINW, NOUT, respectively.

We now test whether there are correlations:

- between the technological profile of the domestic base of a country and that of foreign subsidiaries in that country [correlation between vectors nDOM(i) and nINW(i)], on the one hand;
- between the technological profile of the domestic base of a country and that of subsidiaries in foreign countries of the MNEs of that country [correlation between vectors nDOM(i) and nOUT(i)], on the other.

In order to test these hypotheses statistically, we estimate the parameters of a linear regression of nDOM(i,k) on nINW(i,k) and on nOUT(i,k). For each country (i), we have the following equations:

$$nDOM(i,k) = a(i) + a^{*}(i)\, nINW(i,k) + e(i,k) \quad [k = 1, 2,...6] \quad (1)$$

$$nDOM(i,k) = b(i) + b^{*}(i)\, nOUT(i,k) + é(i,k) \quad [k = 1, 2,...6] \quad (2)$$

where k is the technological area.

a(i), a*(i): regression coefficients of nINW(i,k) on nDOM(i,k);

b(i), b*(i): regression coefficients of nOUT(i,k) on nDOM(i,k);

e(i,k), é(i,k): specification errors on models (1) and (2).

Having computed the coefficients a* and b* at the 95 per cent probability confidence level, we have the following three possibilities :

- they are significantly different from zero and positive: the profiles nDOM and nINW (or nOUT) have a significant positive correlation and thus they exhibit a similar technological orientation;
- they are significantly different from zero and negative: the profiles nDOM and/or nINW (or nOUT) have a significant negative correlation and thus exhibit an opposite technological orientation;
- they are not significantly different from zero: the profiles are unrelated and the technological orientations have no linkage.

Results per country

As Table 9 shows, in five cases out twelve (France, Germany, Switzerland, Japan, EU), there are no correlations between NSIs and inward IIN specialisation. This means that MNEs have a strategy A (local adaptation) or modest-scale strategy B (host-country centralisation).

Table 9. **Characteristics of volume and orientation of the IINs of 11 countries considered as both host and home countries**

	Countries as **host** of subsidiaries of foreign MNEs		Countries as **home** of MNEs with affiliates abroad		Balance of volumes[5]
	Volume[1]	Correlation of specialisations[2]	Volume[3]	Correlation of specialisations[4]	
France	+	none	–	+	+
Germany	–	none	–	none	0
United Kingdom	++	++	+	+	+
Italy	+	++	–	none	++
Netherlands	++	+	++	++	–
Belgium	++	++	+	+	++
Switzerland	+	none	++	++	–
Sweden	+	+	+	++	0
Canada	++	++	+	+	++
United States	–	+	+	none	–
Japan	–	none	–	+	+
EU	–	none	–	+	+

1. INW-IIN index.
2. Correlation between [nDOM] and [nINW].
3. OUT-IIN index.
4. Correlation between [nDOM] and [nOUT].
5. IIN-B index (a + sign means that INW-IIN is greater than OUT-IIN, meaning that the country is a net "exporter" of technology.

Source: Author.

For four countries (United Kingdom, Italy, Belgium, Canada), the correlation is high; since the volumes of inwards IIN are also high, it can be inferred that these countries are the object of strategy B (host-country centralisation) by foreign MNEs.

For outward IINs, only three countries (Germany, Italy, the United States) do not show a correlation of specialisation with the NSI of the home country. This would mean that the MNEs of these countries have strategy C (division of labour), or a tendency towards industrial diversification abroad associated with strategy B (host country R&D centralisation).

For the Netherlands, Switzerland and Sweden, the strong correlation between outward IINs and NSIs, plus high volume, means that the MNEs of these countries have a strong A plus B strategy. Their IINs are an extension of their NSIs, which are too small to support the needs of their big MNEs. Canada presents a quite similar situation.

Typology of countries based on the characteristics of their inwards and outwards IINs

There are four major types of country profiles in relation to IINs (Table 10):

- Type A (United Kingdom, Canada, Belgium, Italy) has very significant inward IIN; it is so embedded in the NSI that its technological orientation is highly correlated with it. The volume of inwards IIN is much higher than that of outward IIN, indicating significant exports of technological activities. Sub-types A1 and A2 differ in that the latter makes little use of outward IIN. This situation may raise – for both sub-types – the question of the longer-term internal coherence of the NSI.
- Type B (Netherlands, Switzerland, Sweden has important outward IINs, which constitute an extension of the NSI and thus have a very similar technological orientation. Foreign MNEs have only small IINs in the country. The NSIs of these countries are "bigger" than they appear. In sub-type B2 the outward IINs are less highly developed.
- Type C has very limited (sub-type C1 – Japan) or rather limited (sub-type C2 – Germany, France, the EU) IINs in relation to the size of the NSIs. This type exhibits balanced (Japan and Germany) or moderately over-sized inward IINs (France and EU). The inward IINs are not correlated with NSIs. This tends to show that foreign MNEs have a strategy A (local adaptation) or limited-scale strategy B (home-country centralisation) in these countries.

Table 10. **Country types with respect to the technological strategies of MNEs**

	Countries as **host** of subsidiaries of foreign MNEs		Countries as **home** of MNEs with affiliates abroad		Balance of volumes	Countries
	Volume	Correlation of specialisation	Volume	Correlation of specialisation		
Type A1	++	++	+	+	++	UK, Canada, Belgium
Type A2	++	++	–	+/none	++	Italy
Type B1	+	+/none	++	++	–	Netherlands, Switzerland
Type B2	+	+	+	++	0	Sweden
Type C1	–	none	–	+	+	Japan
Type C2	+/–	none	–	+/none	0/+	Germany, France, EU
Type D	–	+	+	none	–	United States

Source: Author.

- Type D (United States) has limited inward IINs, developed with strategy C (division of labour) by foreign MNEs, and significant outward IINs, developed with strategy A (local adaptation).

CONCLUDING COMMENTS

There are two lines of critical analysis:

- first, any measurement of IINs can be challenged on two grounds: we assume that the country of the patent inventor is a valid proxy for technological innovation activity in that country and that financial linkages among firms can be channels for technological linkages;
- second, the interpretation of the technological orientations (correlations) can be challenged: what we measure is an aggregate result of the strategies of all MNEs of a country or regarding a country and we do not know the situation for specific firms and technologies, which may well differ.

Nevertheless, the results presented here point to the importance and specificities of the two-way relationships between IINs and NSIs. The way NSIs are related to the IINs of MNEs differs widely among countries and is an important characteristic of the NSI. We suggest that the dynamism of NSIs is, in some cases, heavily influenced by the specifities of their international insertion.

Further work should overcome these limitations and also cover the international scientific networks linking MNEs and NSIs and NSIs with each other.

We agree with the results of Patel and Pavitt (1991) and Patel (1995) on the importance of the home-based system of innovation for the performance of MNEs, but we consider that this does not imply that IINs can be treated as negligible.

NOTES

1. The data on MNEs' European patenting was prepared by Françoise Laville (OST) and Jean-Pierre Bordet (ASSI); the statistical tests were performed by Régis N'Sondé (Université Paris-IX) and Catherine Carpentier (OST).
2. For a complete overview, see OECD (1992) and OECD (1994).
3. This term is used in some of the litterature in English; in French, the OECD uses the same word, but others prefer "mondialisation". In all cases, the idea is to go beyond the word "internationalisation" in order to make clear that the new quantitative scale of the phenomenon has meant a qualitative difference. Here, we will use the word "internationalisation" while making explicit what it means.
4. For a synthesis, see, for example, Muchielli (1985).
5. The IIN of the MNE also involves linkages with university research, be it national or located in other countries. An appraisal of international networks between basic research and MNEs is proposed in Zitt and Barré (1994). Another approach can be found in Hicks *et al.* (1994).
6. In what follows, an MNE is considered to be constituted first of a "firm of origin", its country being the "home country" of the MNE, and second of "subsidiaries", their country of localisation – where "local" operations take place – being the "host country".
7. The studies include case studies as well as small sample (non-representative) quantitative analyses; Patel and Pavitt, who base their analysis on thorough quantitative analysis, have primarily investigated the impacts of NSIs on IINs and less the reverse (feed-back) effects.
8. Beyond this basic formulation, the question of the "nationality" of firms arises: who benefits from the oligopoly rent? This leads to the question of whether a technology developped in one country, in one particular subsidiary, can be exploited in another country (see, for example, Brainard, 1993).
9. For a review, see Granstrand *et al.* (1993); for an additional example, see Papanastassiou and Pearce (1993).
10. For European countries, the systematic innovation surveys, which have some questions about IINs, may produce a new situation; nevertheless, the quasi-impossibility for researchers to access the data makes it impossible to consider this alternative seriously.
11. The various databases on R&D expenditures at the level of the world's major firms (for example the Data for European Business Analysis (DEBA) database of the European Commission are already consolidated, without a breakdown by subsidiary and country of origin. Worse, the subsidiaries considered are usually not listed. Therefore, these databases are of very little use indeed for the problem raised here.

12. Studies using the patent-consolidation method have also recently been carried out for the European Commission by various economists.

13. The best-known is the one produced by MERIT.

14. On this newly developing indicator, see Callon and Carpentiers (1994).

15. The data exist at the level of 30 technological sub-areas, but are not presented here; only the analysis of the technological orientations is based on the 30 sub-areas.

16. The database itself was developed by Françoise Laville.

17. For a detailed definition, see OST (1993), Annex A-6.

18. For example, the patent's world share of country i in technological area k.

REFERENCES

ARCHIBUGI, D. and M. PIANTA (1992), "Specialization and Size of Technological Activities in Industrial Countries: The Analysis of Patent Data", *Research Policy*, Vol. 21, pp. 79-94.

BRAINARD, R. (1993), "Globalisation and Corporate Nationality", *STI Revue*, No. 13, OECD, Paris, pp. 163-190.

CALLON, M. and CARPENTIER, C. (1994), "Co-patenting as an Indicator of Socio-economic Networks", paper presented at the EUNETICS Conference, Strasburg, October.

CANTWELL, J. (1989), *Technological Innovation and Multinational Corporations*, Blackwell, London.

CANTWELL, J. (1991), "La théorie de la compétence technologique et son application à la production internationale", in D. McFetridge (ed.), *Investissement étranger, technologie et croissance économique*, pp. 41-84, The University of Calgary Press, Calgary.

CANTWELL, J. and HODSON, C. (1990), "The Internationalization of Technological Activity and British Competitiveness: A Review of Some New Evidence", in Cason, M.C. (ed.), *Global research strategy and international competitiveness*, Basil Blackwell, Oxford.

DOSI, G., K. PAVITT and L. SOETE (1990), *The Economics of Technical Change and International Trade*, Harvester Wheatsheaf, London.

DUNNING, J.H. (1994), "Multinational Enterprises and the Globalization of Innovatory Capacity", *Research Policy*, Vol. 23, pp. 67-88.

GRANSTRAND, O., L. HAKANSON and S. SJOLANDER (1993), "Internationalization of R&D – A Survey of some Recent Research", *Research Policy*, Vol. 22, pp. 413-430.

HAGEDOORN, J. and J. SHAKENRAD (1992), "Leading Companies and Networks of Strategic Alliances in Information Technologies', *Research Policy*, Vol. 21, pp. 163-190.

HARRIS, R. G. (1991), "Politique commerciale stratégique, retombées technologiques et investissement étranger", in D. McFetridge (ed.), *Investissement étranger, technologie et croissance économique*, The University of Calgary Press, Calgary.

HICKS, D., T. ISHIKUBA, P. KEEN and S. SWEET (1994), "Japanese Corporations, Scientific Research and Globalization", *Research Policy*, Vol. 23, pp. 375-384.

HOWELLS, J. and M. WOOD (1993), *The Globalization of Production and Technology*, Belhaven, London.

MADEUF, B. *et al.* (1993), "Les activités de recherche en France des sociétés étrangères", DGRT, ministère de la Recherche et de l'Enseignement supérieur, October, mimeo.

MAARTEN DE VIET, J. (1993), "Globalisation and Local and Regional Competitiveness", *STI Revue*, No. 13, OECD, Paris, pp. 89-122.

MUCHIELLI, J.L. (1985), *Les firmes multinationales : mutations et nouvelles perspectives*, Economica, Paris.

NATIONAL SCIENCE BOARD (1994), *Science and Engineering Indicators*, USGPO, Washington, DC.

OECD (1992), *Technology and the Economy: The Key Relationships*, Chapter 10, Paris.

OECD (1993), *Globalisation and Local and Regional Competitiveness*, OECD Working Documents, II,16, Paris.

OECD (1994), *Globalisation of Industrial Activities*, Paris.

OST (1993), "Science et technologie – indicateurs 1994 ; Observatoire des Sciences et des Techniques", *Economica*, Paris.

PAPANASTASSIOU, M. and R. PEARCE (1993), "La mondialisation de l'innovation et l'organisation de la RD dans les multinationales", in F. Sachwald (ed.), *L'Europe et la globalisation – acquisitions et accords dans l'industrie*, pp. 69-106, IFRI-Masson, Paris.

PATEL, P. (1995), "Localized Production of Technology for Global Markets", *Cambridge Journal of Economics*, Vol. 19, pp. 141-153.

PATEL, P. and K. PAVITT (1991), "Large Firms in the Production of the World's Technology: An Important Case of Non-globalization", *Journal of International Business Studies*, Vol. 22, No. 1, pp. 1-22.

PAVITT, K. and P. PATEL (1991), "Technological Strategies of the World's Largest Companies", *Science and Public Policy*, Vol. 18, No. 6, pp. 363-368.

SACHWALD, F. (1993), "Mondialisation et systèmes nationaux", in F. Sachwald (ed.), *L'Europe et la globalisation – acquisitions et accords dans l'industrie*, pp. 15-67, IFRI-Masson, Paris.

ZITT, M. and R. BARRÉ (1994), "Positioning Actors in the Interacting Science-technology System by Citation Flow Analysis", paper presented at the EUNETICS Conference, Strasbourg, October.

INTERNATIONAL PATENTING STRATEGIES OF MULTINATIONAL CONCERNS: THE EXAMPLE OF TELECOMMUNICATIONS MANUFACTURERS*

by
Ulrich Schmoch
Fraunhofer Institute for Systems and Innovation Research, Karlsruhe, Germany

INTRODUCTION

Patents are valuable indicators of the international diffusion of innovations. Thus, among Western Europe, North America and Japan, patent flows reflect flows of technology. It is, however, quite difficult to interpret the figures adequately, because special features of different regional situations influence the outcome. Nevertheless, using the example of selected telecommunication manufacturers, it is possible to establish linkages between international patenting strategies and international technology strategies. A comparison of the external patenting patterns of certain companies to average patterns of foreign patenting makes it possible to draw at least qualitative conclusions about individual international market strategies. A major result of such an analysis is that Japanese, North American and European companies generally pursue different international strategies. For example, US-based companies are generally more interested in the Japanese market than European ones and Japanese companies are more selective than American or European ones.

Technological change and innovation are important factors for productivity and international competitiveness and have thus become a central topic of economic analysis in most industrialised countries. The major concern of analysts is to describe science and technology (S&T) activities in quantitative as well as qualitative terms. However, because S&T can only be measured indirectly using different types of indicators, patents have proved to be valuable indicators for describing industrial technological activities, provided that they are used with sufficient methodological care (OECD, 1994). Among their other advantages, patents have a very detailed technical classification, which allows almost unlimited choice of

* The paper is based on two studies for the OECD, Paris and the Wissenschaftliches Institut für Kommunikations Clienste (WIK), *Bad Honnef. The author is grateful to his colleague Norbert Kirsch and to Dr. Thomas Schnöring, formerly of WIK and now at DBP Telekom, for many fruitful discussions.*

aggregation levels from broad fields of technology down to single products. In this respect, patent indicators are much better than R&D expenditure or trade and production statistics, where the degree of disaggregation is much less.

The most common types of patent indicators use the number of patents at representative patent offices, such as the US Patent and Trademark Office (USPTO) or the European Patent Office (EPO), broken down by countries, companies, or technological fields. They are used to measure technological performance or to describe technological profiles or specialisations. In addition, counts of patent citations are used for more sophisticated analyses of technological performance or technological linkage. As part of the discussion of international competitiveness, globalisation of technology – the so-called techno-globalism – plays a major role (*e.g.* Dunning, 1992; Patel and Pavitt, 1990; or Pearce and Singh, 1992). The globalisation of technology has different dimensions, of which the most important are:

- global exploitation of nationally generated technology through international trade;
- global technological collaboration by companies from different countries;
- global generation of technology through international R&D of multinational companies.

In addition to these three categories proposed by Archibugi and Michie (1995), global diffusion of knowledge within multinational companies due to foreign production should be taken into account, because it may occur even without foreign R&D.

Patent indicators may be used to analyse technoglobalism at the country level as well as for single companies. The following discussion addresses the question of whether patent indicators are really an appropriate tool for describing this phenomenon. It focuses on global exploitation or diffusion of technology and, in addition, briefly considers global generation of technology. The question of global technological collaboration is investigated in detail in Schmoch *et al.* (1991) and Carpentier and Callon (1994), and global generation of technology in Schmoch and Schnöring (1994).

The analysis starts with general considerations on patent flows within the triad and at country level. Thereafter, international patenting strategies of multinational concerns are analysed, using selected telecommunications manufacturers as an example.

INTERNATIONAL PATENT FLOWS BETWEEN COUNTRIES

To understand the meaning of international patenting, it is necessary to explain at least some basic elements of the international patent system; a more detailed description is given in Schmoch (1990) or OECD (1994). Under the present legal system, patent protection is only valid for national territories. The EPO has a centralised application and examination procedure, but the European patent, once granted, is only valid for certain countries. For international applications in the framework of the Patent Co-operation Treaty, applications are centralised, but the preliminary examination is not legally binding on the destination countries. In any case, a company has to register an invention in every country where it seeks patent protection – and, where appropriate, with an international or a European application as intermediate steps.

Patents are industrial property rights and are always closely linked to commercial strategies. Apart from the main criteria of novelty and inventive step, patents must be "susceptible of industrial application" (see, for example, Article 52 of the European Patent Convention). Furthermore, patent protection is very expensive. Debontride (1993) reports the example of a French invention, which, according to experts, seems to be representative of the average costs of patents. The cost of the domestic application in France, including the fee of the patent attorney and the official fees for the application and for five years' renewal amounted to US$4 800. The costs for the parallel European application in seven countries, up to and including final national registrations, amounted to nearly $40 000. The cost for applications in Japan and the United States came to about $7 000 and $8 000, respectively. These figures clearly demonstrate that patent applications, especially in foreign countries, are only made if there is a clear market interest. Against this background, patent registration in foreign countries is obviously an appropriate indicator of external market strategies.

The legal protection offered by patents covers, on the one hand, commercialisation – the offer, sale or import of the patented product or process – and, on the other, its production. Therefore, the goal of a patent registration may be to protect a new good introduced into a special national market; in this case, the patent strategy is an offensive one. A further goal in filing a patent may be to prevent a competitor from producing a new product or implementing a new process; in this case, the strategy is defensive. In most cases, it is quite difficult to differentiate clearly between these two goals. There are, of course, other reasons for patenting, for example, to build up a portfolio for future licencing or technology exchange agreements with competitors (Bertin, 1992, p. 91).

Against this background, it is possible to put forward initial qualitative interpretations of international patent flows among Western Europe, North America and Japan. The following considerations concern the first year of patent registration, the so-called priority year, which is nearly the same as that of the invention. The national origin of the patent is defined by the country of the inventor; this is the most appropriate way to represent where the technology originated (see Schmoch and Kirsch, 1994). North American patent applications are represented by patents at the USPTO, because almost all foreign applications in Canada are registered in the

United States as well. In order to achieve comparable figures for Europe and Japan, the US figures are also based on applications, although the USPTO only officially publishes information on patents granted. When comparing the statistics given below with official data from national patent offices or the World Intellectual Property Organisation (WIPO), it should be noted that not all officially registered applications, but only those published, are taken into account, because only these documents are available in on-line databases.

Figure 1 represents the flows of patent applications within the triad for 1989. It shows that most Western European inventions registered in North America come from the European Union (EU) (25 721 out of 30 028, or 86 per cent). In the opposite direction, all patents from North America to Western Europe are registered in at least one EU country, so that it is not necessary to distinguish between Western Europe and the EU. Flows of patent applications between North America and the EU are nearly balanced, with 22 321 applications from North America and 25 721 applications from the EU.

Only 44 per cent of the European (and EU) applications in North America result in applications in Japan as well; this can be taken as a clear indication of limited interest in the Japanese market. Similarly, the share of North American inventions registered in Japan was only 56 per cent. It is unreasonable to add up registrations in different European countries, because this would lead to an enormous overrating of patent activities as compared to Japan or North America (Schmoch *et al.*, 1988;

◆ Figure 1. ***General flows of patent applications between triad regions for the priority year 1989***

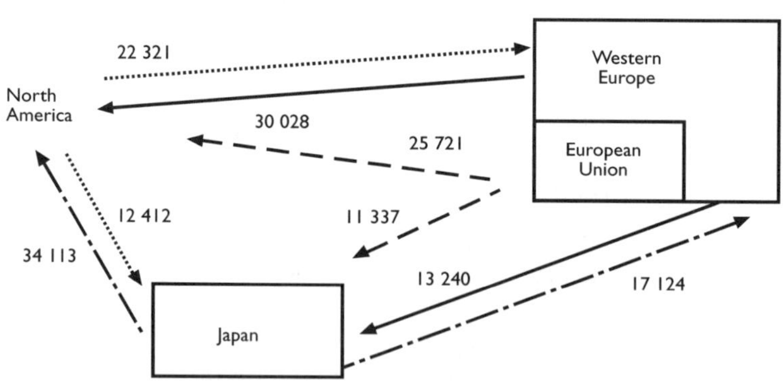

Source: Schmoch and Kirsch, 1994.

Mogee and Kolar, 1993). In consequence, the following statistics are based on patent families, defined as a group of parallel patent applications linked to one original invention. For example, an invention of American origin is counted when it is registered in at least one European country.

From the Japanese perspective, a flow of 34 113 applications documents a high interest in the North American market. It is larger than the flow of all European countries to North America and much larger than the flow from North America to Japan. Only 50 per cent of the Japanese applications in North America are registered in Western Europe as well; this indicates a clear Japanese preference for the North American market. This preference can be explained by the shorter distance from Japan to North America than to Western Europe, but especially by the very large US market, whereas – at least in 1989 – the European market was still quite fragmented, and special conditions had to be observed in each country. The increasing unification of the European market has led to growing trade flows within the European Union (Grupp and Soete, 1992), and it can be assumed that the interest of non-European countries – including Japan – in this market will grow as well.

A comparison of patent application flows with trade flows illustrates their close relation. Taking Germany as a core country of Western Europe, one finds that imports and exports between Germany and the United States are nearly balanced; in contrast, German imports from Japan are about three times exports, a ratio also found for patent application flows between Germany and Japan (Schmoch and Kirsch, 1994; BMFT, 1993, p. 21).

As international flows of patent applications obviously reflect international technology flows in terms of international technology trade, patent statistics can be used to test the thesis of growing technoglobalism. Figure 2 illustrates the situation on the large and representative US market; it show an enormous increase in the number of foreign patents, which were multipled by 1.75 between 1980 and 1990, *i.e.* 5.7 per cent a year (Schmoch and Kirsch, 1994). This is largely due to the increase in Japanese patenting in the United States, which grew at an annual rate of 9.5 per cent. Even without patent applications of Japanese origin, however, foreign patents grew by 2.7 per cent annually. The number of foreign patents on the Japanese market has increased as well, but to a lesser degree; the number of foreign applications increased by hardly 1 per cent a year. All in all, the patent analysis confirms the assumption of a growing globalisation of technology, but the change in the 1980s is relatively slow, with the major exception of technology flows from Japan.

The above considerations for the triad indicate that international patent flows can only be adequately interpreted by taking into account the geographical location and the size of the country of origin and of the respective external markets. Furthermore, it is necessary to take account of the impact of different legislative regimes

◆ Figure 2. ***Foreign patent applications in the United States and Japan***

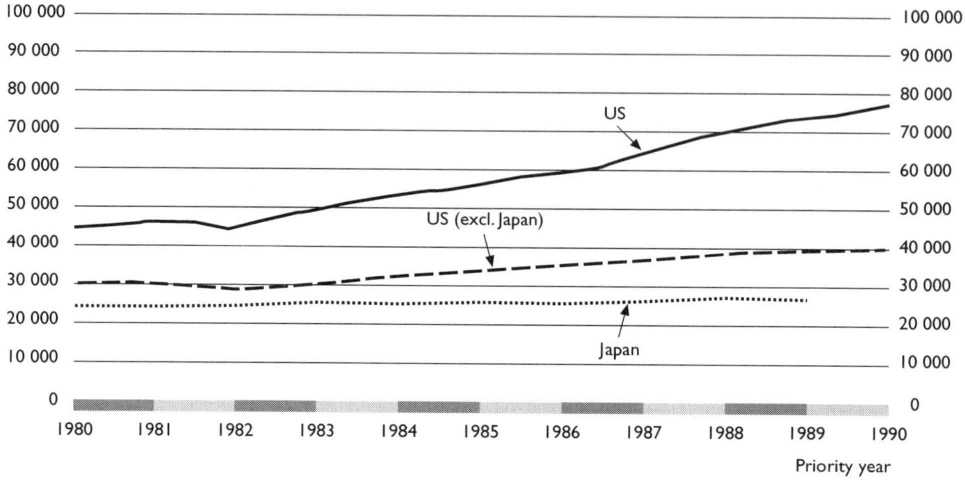

Source: Schmoch and Kirsch, 1994.

(Auriol and Pham, 1993). Against this background, interpreting general figures on foreign patenting, such as the total number of non-resident patents (WIPO, 1992) or a certain country's share of all foreign patents (Vinkler, 1993), is quite dangerous, because the interests of completely different countries are mingled. For example, the interests of Japan, France and Canada in the American market are quite different, owing to their different geographical situations. As a consequence, it is necessary to examine countries of origin individually.

Figure 3 shows the pattern of patent application flows for the United States. As the US market is the world's largest national market, the inventions of US inventors are mainly registered in the United States. Only 26 per cent of domestic applications are filed in Western Europe, and only 15 per cent in Japan. For Western Europe, it is interesting to look as well at the number of applications to the EPO, as these generally concern more than three European destination countries. In consequence, a high share of EPO applications imply broad coverage of the European market. For US applications in Western Europe, 77 per cent are filed through the EPO.

Table 1 documents patent flows for France, Austria and Japan in order to gain a reference point for the US situation. It gives an index value of 1 for the largest patent flow to a different region. Thus, for the United States, the patent flow to

◆ Figure 3. ***Flows of patent applications of US origin for the priority year 1989***

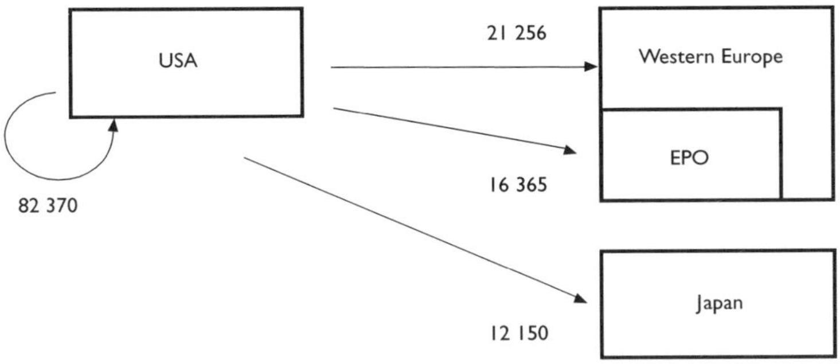

Source: Schmoch and Kirsch, 1994.

Table 1. **Flows of patent applications from selected countries for the priority year 1989**

Origin	Destination				
	Domestic	North America	Western Europe	EPO	Japan
United States	3.88	–	1.00	0.77	0.57
France	2.02	0.81	1.00	0.87	0.35
Austria	1.25	0.53	1.00	0.66	0.17
Japan	8.87	1.00	0.50	0.37	–

Source: Schmoch and Kirsch, 1994.

Western Europe receives a value of 1 and the domestic patents receive a value of 3.88. France represents the situation of a large European country, most of whose external patents are registered in neighbouring Western European countries. As the French market is smaller than the American one, the relative share of domestic patents is much lower than in the United States. Within Europe, the relative share of patents filed through the EPO is higher than that of American applications, thereby documenting broader coverage of the European market. In contrast, the relative flow of patents to the Japanese market is distinctly lower than for the United States.

Austria is a small European country; in 1989, it was not yet a member of the European Union. The largest flow of external patents naturally went to Western Europe, but the relative share of EPO applications is lower than for France. This is due to the fact that a large part of foreign patents were registered exclusively in Germany. It is not surprising that the relative share of Austria's domestic applications is only slightly higher than the number of foreign patents and in consequence distinctly lower than in France. The quite low share of Austrian patents in North America and Japan illustrates the difficulty for a small country to enter remote overseas markets.

Japan has already been discussed in the context of triad flows. Table 1 also shows the strikingly high number of domestic patents, which is due to Japan's very special patent "culture". Until recently, Japanese patent legislation was based on the so-called single claim system, which meant that several patents were required to achieve the same protection as a single patent in the United States or Europe (Auriol and Pham, 1993). Although the present patent law allows for multi-claim applications, the high propensity to patents has changed little. Furthermore, patents are generally highly esteemed as an instrument for promoting innovation, are important for the standing of companies, and are crucial for the career of scientists and engineers (OECD, 1994, p. 22). In this context, two observations should be made. First, the relation between domestic and foreign patents of Japanese origin shows that, owing to the high costs of foreign patenting, the level of foreign patents is "normal" in comparison to other countries. Second, foreign applicants in Japan took little notice of the single-claim system and simply registered the core idea of their original domestic patents. In consequence, the number of foreign patents in Japan remained quite low, primarily because of limited interest in the Japanese market.

The observation that legislative regimes have little impact on the volume of international flows holds for other countries as well. For example, Europeans registering a patent in the United States adapt their applications to the US legal interpretation of claims, but this does not significantly affect the number of applications. The differences in the systems are only important for patent statistics, because for Europe and Japan, published applications are available in on-line databases, whereas the United States only publishes patents granted. Only in Central and Eastern European countries do legislative regimes have a marked effect on foreign patenting, because until around 1989 these were socialist economies and patents had a completely different meaning than in Western countries. At present, these countries are slowly adapting to Western patent systems.

With knowledge of the situation of individual countries, it is possible to interpret general patent indicators, such as the average number of foreign patents for patent families with at least one foreign application, in an adequate way. Table 2 shows the values for the United States, Japan and Germany for 1972 and 1984. The

Table 2. **Size of patent families with at least one foreign application for selected countries**

Priority year	Country of origin		
	United States	Germany	Japan
1972	4.8	4.7	2.9
1984	7.1	6.7	3.6

Source: Faust, 1987; cited in Schmoch *et al.*, 1988, p. 93.

comparison shows a considerable growth in foreign patenting for all countries and thus confirms the technoglobalisation thesis, especially for the 1970s. The relatively low values for Japan are due to a clear preference for the American market, whereas the broad coverage of European countries by German and American companies results in higher values.

INTERNATIONAL PATENT STRATEGIES OF SELECTED COMPANIES

On the basis of observations of the general patterns of international patent flows, it is possible to interpret the international patenting strategies of selected companies. To this end, the telecommunications-related patent applications of a sample of 21 telecommunications companies were analysed for the priority period 1987-89 for a selected set of destination countries. In telecommunications, R&D intensity is very high, at 5 to 12 per cent of turnover for R&D expenditure, and patents therefore play a major role in protecting investment in innovations at the national and international levels. It should be noted that – for switching technology, for example – even software-related inventions can be and are registered as patents, because they are closely linked to improvement of hardware.

As for the interpretation of patent flows at country level, evaluation of company strategies requires further background information, such as parent country, type of company, etc. Figure 4 shows the relative distribution of the destination countries for Siemens; the country with the highest number of foreign applications, the United Kingdom, is given the index value 1 (Schmoch and Schnöring, 1994). On this basis, the figure for domestic applications in Germany is only about 1.7. This figure is low compared to the average situation, in which only about one-third of German domestic applications are registered in one or more foreign countries as well. This finding confirms that the high value placed on registered inventions in the telecommunications area leads to a high share of foreign patent applications. All Siemens' foreign patents in Europe are registered in both France and the United Kingdom, and Italy and the Netherlands also have high shares (indices of 0.9 and 0.8,

◆ Figure 4. ***Destination countries of telecommunications-related patent applications by Siemens, 1987-89***

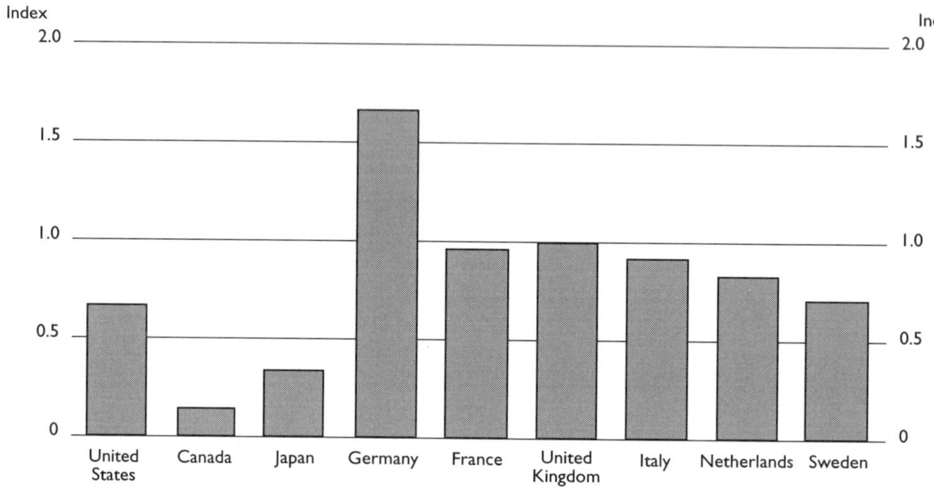

Source: Schmoch and Schnöring, 1994.

respectively. Determining the number of patents at the USPTO raises methodological problems. As already noted, only patents granted are published, so that it is necessary to estimate applications from average granting rates. Second, as the data set for 1987-89 had not been completed in spring 1992 when the database searches were made, the US values have a certain margin of error. On the basis of these estimates, about 70 per cent of Siemens' European patents are also registered in the United States but only 35 per cent in Japan. Table 3 gives a general overview of the international patenting strategies of the selected companies. For Europe, it is obvious that, compared to France and the United Kingdom, Germany's share does not reflect market volumes. This is due to the fact that patent registration is a simple yes-or-no decision. Therefore, there is a certain differentiation between larger and smaller European countries, but not among larger countries. In the smaller countries, the wish to block a major competitor often plays a decisive role; for example, registrations in the Netherlands and Sweden have to be seen in the context of Philips and Ericsson.

As it is not possible to discuss here the strategies of all selected companies in detail, typical strategies are summarised in Table 4. A first group of companies has an average share of foreign patents; the average is defined by the general situation in telecommunications. If these companies go into foreign markets, they patent widely. Large companies, such as Alcatel, AT&T, Philips and Siemens belong to this

Table 3. **Destination countries for telecommunications-related patent applications for selected companies, 1987-89**[1]

Concern	Destination country								
	US	CAN	JP	DE	FR	GB	IT	NL	SE
Alcatel NV	0.71	0.71	0.49	1.28	1.09	1	0.97	0.92	0.89
Bosch	0.4	0.11	0.39	1.98	1	0.95	0.92	0.69	0.58
Ericsson	0.88	0.37	0.71	0.92	0.85	1	0.69	0.76	1.02
GEC	0.66	0.37	0.73	1	0.99	1.3	0.91	0.81	0.81
STET/Italtel	0.63	0.34	0.5	1	1	1	0.89	1	0.97
Nokia	0.28	0.06	0.32	0.95	0.94	1	0.81	0.71	0.82
Philips	0.92	0.19	0.85	1.1	0.98	1	0.69	0.44	0.59
Siemens	0.67	0.14	0.34	1.68	0.97	1	0.92	0.83	0.71
Thomson	0.97	0.13	0.48	0.99	1.23	1	0.76	0.52	0.43
AT&T	1.96	0.84	0.98	0.93	0.93	1	0.65	0.54	0.51
IBM	1.11	0.1	0.85	1	1	1	0.32	0.1	0.08
Motorola	1.51	0.44	0.94	0.95	0.95	1	0.9	0.92	0.9
Northern Telecom	1	0.48	0.38	0.36	0.36	0.38	0.2	0.35	0.34
Fujitsu	0.9	0.69	31.46	1	0.93	0.99	0.25	0.11	0.23
Hitachi	1	0.11	12.19	0.54	0.33	0.39	0.07	0.06	0.05
Matsushita	1	0.12	3.2	0.66	0.53	0.7	0.13	0.27	0.26
NEC	1	0.5	3.81	0.58	0.43	0.69	0.11	0.29	0.26
NTT	1	0.5	44.8	0.87	0.51	0.79	0.14	0.27	0.52
OKI	1	0.19	20.95	0.53	0.47	0.57	0.15	0.04	0.19
Sony	1	0.27	5.25	0.85	0.78	0.89	0.11	0.34	0.04
Toshiba	1	0.32	13.34	0.54	0.33	0.47	0.05	0.08	0.13

1. US = United States; CA = Canada; JP = Japan; DE = Germany; FR = France; GB = Great Britain; IT = Italy; NL = Netherlands; SE = Sweden.

Source: Schmoch *et al.*, 1992.

Table 4. **Typical patent strategies of selected companies on foreign markets**

Feature	Examples
Average share of foreign patents, broad coverage	Alcatel, AT&T, Philips, Siemens
Generally, little foreign patenting, but broad coverage	Ericsson, GEC, Motorola, many network operators
Selective strategy	Hitachi, Oki, Matsushita, NorTel
Special focus on the American market	Matsushita, Hitachi, NEC, Thomson, Philips
Special focus on the Japanese market	AT&T, IBM, Motorola, NorTel, Philips
Low presence on the Japanese market	Bosch, Nokia, Siemens, STET, Thomson

Source: Schmoch and Schnöring, 1994.

group. A second group has a quite restrictive patent policy: the number of foreign, but also domestic, patents is low compared to the related R&D budget (Schmoch and Schnöring, 1994). If these companies decide to register an invention, they cover a large number of foreign countries. This indicates the high value of the few inventions that are patented. Typical companies in this group are Ericsson, GEC, Motorola and also network operators such as NTT. In the original report of Schmoch *et al.* (1992), the strategy of other network operators such as British Telecom, France Télécom, Bellcore and GTE are also discussed and show similar patterns.

A characteristic of most Japanese companies, but also of Northern Telecom, is a very selective patent strategy in foreign countries. Japanese companies choose first the US and two or three large European markets, but generally neglect the smaller European ones. This indicates that Japanese companies are very interested in the US market, but that since they are – not yet – well represented in Europe, a broader protection strategy is not reasonable. Fujitsu is a major exception, as it has a comparable share of foreign patenting in the United States and the three major European countries. This difference may be due to the specific technological strategy of this company which is focused on line-bound transmission, *i.e.* electrical transmission, optical transmission and switching technology (Schmoch and Schnöring, 1994, p. 403), whereas the focus of the other Japanese companies in the sample is on terminals and radio transmission. This thesis would, of course, have to be confirmed by further investigations such as interviews. The patent analysis, however, helps to ask the appropriate questions for such types of in-depth analysis.

As for the general figures at the country level, Japanese companies have an enormous number of domestic patents, from five to 40 times higher than the maximum number of patents for a foreign market. These values are generally higher than the overall figures in Table 1 and indicate a concentration on the Japanese market. On the basis of present knowledge, it seems impossible to draw further conclusions from internal differences. For NTT, however, the enormous difference between domestic and foreign patenting can be explained by its role as major network operator in Japan, so that external patenting is less important.

As already mentioned, most Japanese companies focus strongly on the US market. On the European side, only Thomson and Philips show a similar strong orientation. The North American companies AT&T, IBM, Motorola and Northern Telecom have quite high patent numbers in Japan, whereas most European companies have low ones; only Philips is well represented in Japan. This does not mean that European companies should protect their inventions in the Japanese market to a greater extent. If the market potential is insufficient, it is not reasonable to make expensive patent applications. Nevertheless, these findings indicate that European companies have not yet established a sufficient presence in the Japanese market, a key market for access to the whole East Asian region.

These international patent strategies correspond to international trade patterns. Traditionally, the US market has been far more important for Japanese telecommunications equipment exports than European ones, although some changes have occurred during the 1980s (Schmoch and Schnöring, 1994, p. 410). The Japanese market is of growing importance for US exports, but is still totally unimportant for the exports of EU countries. Japan's imports of telecommunications equipment, mostly from North America, have increased rapidly during recent years, with double-digit growth rates. Motorola, AT&T and Northern Telecom have a foothold in the Japanese market, but it is quite likely that US trade policy has eased this market entry (Schnöring, 1992).

All in all, patent indicators are an appropriate and effective tool for examining the issue of technoglobalism. The analysis, however, has to start with individual countries or companies, because this is the only appropriate way to interpret more general indicators.

REFERENCES

ARCHIBUGI, D. and J. MICHIE (1995), "The Globalisation of Technology: A New Taxonomy", *Cambridge Journal of Economics*, Vol. 19, No. 1, February, pp. 121-140.

AURIOL, L. and F. PHAM. (1993), "What Pattern in Patents", *OECD Observer 179*, December 1992/January 1993, pp. 15-18.

BERTIN, G.Y. (1992), "Patent and Licensing Strategies of the Various Technology Types of Corporations", Proceedings of the First EPO-IFO Workshop, 19-20 March, held at the European Patent Office, *Results and Methods of Economic Patent Research*, pp. 83-101.

BMFT (1993), *Bundesbericht Forschung 1993*, Bonn.

CARPENTIER, C. and M. CALLON (1994), "Cooperation and Technological Dynamics: Patent Co-Filing", *EUNETIC Proceedings*, Vol. 1, Strasbourg, 6-8 October, pp. 669-675.

DEBONTRIDE, X. (1993), "Le brevet : une arme stratégique à découvrir", *Usine nouvelle*, No. 2414, June, pp. 36-39.

DUNNING, J. (1992), *Multinational Enterprises and the Global Economy*, Addison-Wesley, Wokingham.

FAUST, K. (1987), *Früherkennung technischer Entwicklungen auf der Basisi von Patentdaten*, IFO-Institut für Wirtschaftsforschung, Munich.

GRUPP, H. and L. SOETE (1992), *Analysis of the Dynamic Relationship between Technical and Economic Performances in Information and Telecommunication Sectors*, Vol. 1: *Synthesis Report*, Karlsruhe, Maastricht.

MOGEE, M.E. and R.G. KOLAR (1993), "International Patent Indicators of US Position in Critical Technologies", Report to the NSF, Great Falls, VA.

OECD (1994), "The Measurement of Scientific and Technological Activities: Using Patent Data as Science and Technology Indicators. Patent Manual 1994", OCDE/GD(94)114, Paris.

PATEL, P. and K. PAVITT (1990), "Large Firms in the Production of World's Technology: An Important Case of Non-globalisation", *International Journal of Business Studies*, first quarter, pp. 1-21.

PEARCE, R. and S. SINGH (1992), *Globalising Research and Development*, London.

SCHMOCH, U. (1990), *Wettbewerbsvorsprung durch Patentinformation*, Cologne.

SCHMOCH, U., H. GRUPP and U. KUNTZE (1991), "Patents as Indicators of the Utility of European Community R&D Programmes", in CEC, *Monitor/Spear Report* EUR 13661 EN, Brussels and Luxembourg.

SCHMOCH, U., H. GRUPP, W. MANNSBART and B. SCHWITALLA (1988), *Technikprognosen mit Patentindikatoren*, Cologne.

SCHMOCH, U. and N. KIRSCH (1994), "Analysis of International Patent Flows", unpublished manuscript, Karlsruhe.

SCHMOCH, U., N. KIRSCH, P. GEORGIEFF, H. GRUPP and G. MÜNT (1992), *Internationaler Vergleich von Forschung und Entwicklung im Telekommunikationsbereich – Aktualisierung 1991/92*, 2 vols., Report to the WIK, Karlsruhe.

SCHMOCH, U. and T. SCHNÖRING (1994), "Technological strategies of telecommunications equipment manufacturers – A patent analysis", *Telecommunications Policy*, Vol. 18, No. 5, pp. 397-413.

SCHNÖRING, T. (1992), "Entwicklungstrends auf den europäischen Telekommunikationsmärkten: Auf halbem Weg von monopolitischen nationalen zu wettbewerblichen internationalen Telekommunikationsmärkten", Diskussionsbeiträge No. 102, WIK, Bad Honnef.

VINKLER, P. (1993), "Percentage Patent Representation (PPR), Bilateral Patent Balance (BPB) and Patent Dominancy (PD) Indicators Characterizing International Patenting Relations", *Scientometrics*, Vol. 27, No. 1, pp. 97-103.

WIPO (1992), *Industrial Property Statistics 1990*, Part I: *Patents*, Geneva.

11

PATENTS WITH MULTIPLE INVENTORS RESIDING IN DIFFERENT COUNTRIES[1]

by

William H. Brown and M. James Hirabayashi
US Patent and Trademark Office

INTRODUCTION

Successful development and marketing of technology in today's highly competitive environment often requires the ability to overcome significant challenges, not the least of which are high development costs, the need for specialised expertise, shortened product life cycles, and international competition. In response, government and industry have relied on various types of collaborative efforts to develop and promote technology. Many studies in recent years have documented the proliferation of technology transfer programmes between government sponsors, research institutions, and private industry, as well as the use of joint ventures, licence agreements, acquisitions, mergers, and other co-operative agreements. Such arrangements have fostered the development of technology both domestically and internationally. This chapter concerns "international" collaboration in patented technology, and, in particular, characteristics of patents granted by the US Patent and Trademark Office (USPTO), which have both a US resident inventor and a non-US resident inventor.

Between 1980 and 1993, the number of US patents involving collaborative efforts by US and foreign resident inventors increased from less than 90 to nearly 1 500 patents per year. Data on US patents with at least one US resident inventor and one foreign resident inventor have been analysed by country of inventor residence, patent ownership category, and field of technology. The results of the study document the substantial growth in patent collaboration between US and foreign resident inventors. Technologies emphasized in this group of patents relate primarily to the broad Standard Industrial Classification (SIC) product field of chemicals and allied products (SIC 28) and to chemical technologies in the US Patent Classification System. Analysis of patent ownership illustrates the important role of subsidiaries in the composition of international teams of inventors; however, the analysis is complicated by methodological issues and data anomalies which are discussed in detail.

The growing trend towards international collaboration in scientific research has been documented through the use of patent and non-patent indicators. Between the four-year periods of 1977-80 and 1985-88, for example, the number of international multi-firm research and development alliances grew from 86 to 988; many of these alliances were among firms involved in high-technology areas such as biotechnology and information technologies (Hagedoorn, 1990; cited by National Science Board, 1993). In addition, analysis of scientific and technical articles published during this period shows a sustained increase in the percentage of internationally co-authored articles, from 4.3 per cent in 1977 to 9.1 per cent in 1989 and 11 per cent in 1991 (National Science Board, 1993).

The academic community, as well, has experienced growth in international collaboration. Colleges, universities, and public research institutes in the United States have been particularly active in multinational research. This may be due in large part to the accessibility of institutions with pre-eminent bases of scientific knowledge, which give non-US residents the opportunity to have access to scientific and technical expertise abroad (Howells, 1990).

Within academic and professional disciplines, there is growing recognition that characteristics and trends in patent activity can indicate a great deal about the direction of technological progress and about influential social and economic forces. The development in recent years of computerised databases at the USPTO, which make available enormous quantities of data, have facilitated such analyses. The value of patent data is enhanced by the rarity of available time series for other quantitative indicators spanning the entire technological spectrum. Analyses of patent data can be used to reveal the technological strengths and weaknesses of national economies, diagram the flow of technology among countries, identify global networks of firms and inventors, and highlight significant sources of scientific research. In this chapter, patent data have been used to examine technological research activities involving international collaboration.

Patent data have been used to examine the globalisation of technology from a variety of perspectives. Analyses of filings of domestic and foreign applications have provided evidence that certain technologies tend to be emphasized by certain national economies. This specialisation, coupled with the rapid increase in international patenting, points towards a globalisation of technology markets (Archibugi and Pianta, 1992). Analyses undertaken at the level of the individual firm (*i.e.* the patent assignee) suggest that inventive activity is largely performed within the home country of the assignee. In a study of US patenting by 11 countries that account for more than 95 per cent of OECD R&D expenditures, less than 20 per cent of nationally controlled activities were found to have been performed outside the home country (Patel and Pavitt, 1991).

Analysis of inventor information contained in patent records points to the growing general tendency towards more collaborative research efforts, and towards international collaboration in particular. World-wide, patents with multiple inventors increased from 41 per cent in 1975 to 53 per cent in 1991, and the per cent of patents with multiple, multinational inventors increased from 2.1 per cent to 3.8 per cent. (Schott, 1994). This same research found that fields of technology emphasized in international collaboration are predominately chemical technologies, especially biochemistry, genetic engineering, and organic chemistry. Patterns of collaboration were found to consist of a centre-periphery network with the United States dominating the network. Collaboration has increased among nations of the European Union, but decreased among Eastern European nations, possibly as a result of the increasing complexity of inventive activity and the increasing openness of national societies.

SOURCES OF INFORMATION ON INTERNATIONAL COLLABORATION IN US PATENTS

This study identifies trends in collaboration between US resident and foreign resident inventors, and analyses specifically US patents granted from 1980 to 1993 with at least one inventor who resided in the United States and at least one inventor who resided outside the United States.

The patent data used were derived from two USPTO databases. The first of these, the Technology Assessment and Forecast Program (TAF) database, contains selective bibliographic and status information for US patents issued since 1963. This relational database is most frequently used to report and analyse domestic and foreign patent activity, particularly as it relates to technological development. Reporting is routinely performed through both standardised and custom TAF programme reports. TAF reports and extracts from the TAF database are routinely provided to the public at nominal cost. The second database, the Automated Patent System (APS), contains images of all patents granted since 1790, plus the text-searchable, full text of 1.5 million patents issued since 1971. Although APS is most frequently used by USPTO patent examiners to assist them in their examination of patent applications, APS can be accessed by the public at a reasonable cost from Patent and Trademark Depository Libraries in the United States.[2] The majority of the data used in this analysis were derived from the TAF database. APS data were used to supplement TAF ownership information. The TAF database contains only the first-named patent assignee, while APS contains the names of all assignees that appear on the patent.

The criterion used to identify patents in the TAF database that involve international collaboration between US and foreign inventors required each patent to have at least one US resident inventor and at least one foreign resident inventor. The TAF database contained 10 814 patents granted between 1980-93 that met this criterion. Joint inventors named on US patents are not required to physically work together or work at the same time, nor are they required to make a contribution to the subject matter of every claim of the application. However, each named inventor must have made a contribution, individually or jointly, to the subject matter of at least one claim of the application. Therefore, the criterion used to select these patents is more indicative of international networking in technology development than of the international origin of inventive resources. Patents with US and foreign resident inventors may cover situations in which a foreign inventor moves to the United States and collaborates with an inventor from his or her home country. This might occur when the employee of a foreign firm is reassigned to a US subsidiary. Alternatively, a foreign inventor might relocate in the United States for entirely different reasons.

COLLABORATION BY COUNTRY

The increasing share of patents with US and foreign resident co-inventors provides evidence that pooling of human resources across national boundaries is playing a greater role in obtaining patents at the USPTO. The annual share of patents with US and foreign resident co-inventors grew from 0.2 per cent in 1980 to 2.8 per cent in 1993. This growth in share is notable, particularly in view of the growth in the overall number of patents issued annually during the period. In 1980, 89 patents with US and foreign resident co-inventors were issued, while in 1993, there were 1 499 (Table 1).

US resident inventors were more likely to have foreign resident co-inventors from Canada, Germany, Japan, or the United Kingdom. Over the 1980-93 period, more than 1 400 patents resulted from collaboration with inventors who resided in these countries. While collaboration with Canadian inventors resulted in the greatest number of patents for the entire period (1 781), collaboration between US and Canadian inventors has declined in recent years; only 211 patents were granted in 1993, compared to 279 in 1991. Collaboration with residents of most other countries continued to increase during this period, particularly with German, British, and Japanese resident inventors (Table 2).

Trends in the total number of collaborative patents by country do not necessarily identify which country's residents have the greatest propensity to collaborate with US resident inventors. For example, a large number of the US and foreign resident co-inventor patents include US and Japanese resident co-inventors. However, this number is actually very small when compared to the total volume of patent activity in the United States by Japanese resident inventors.

To identify the extent to which a country's residents emphasize collaboration with US resident inventors, the patents involving US resident and foreign country resident co-inventors were compared with the total number of patents having an inventor from that country. For the 1980-93 period, inventors resident in Israel appear to have a relatively high propensity to co-invent with US resident inventors. Over 11 per cent of the patents with an inventor resident in Israel also name a US resident co-inventor (Table 3). Inventors resident in Japan, in contrast, appear to

Table 1. **Utility patent grants, 1980-93**

Patent counts and percentages by year of grant

Patent category	Year of patent grant														
	1980	1981	1982	1983	1984	1985	1986	1987	1988	1989	1990	1991	1992	1993	Total
Number of patents having a US resident inventor	37 279	39 257	33 879	32 913	38 541	39 808	38 379	43 850	40 724	50 484	47 699	51 583	52 758	53 828	600 982
Number of patents having a US resident inventor and a Foreign resident inventor	89	428	241	346	440	523	575	710	804	1 253	1 116	1 388	1 357	1 499	10 814
% of patents having a US resident inventor which have a foreign resident inventor	0.2	1.1	0.7	1.1	1.1	1.3	1.5	1.6	2.0	2.5	2.4	2.7	2.6	2.8	1.8

Source: USPTO TAF database, 11 July 1994.

Table 2. **Number of utility patent grants with US and foreign resident co-inventors, by country of residence of the foreign resident inventor**[1]

Selected countries

Country of residence of the foreign resident inventor	Year of patent grant														
	1980	1981	1982	1983	1984	1985	1986	1987	1988	1989	1990	1991	1992	1993	Total
Canada	15	71	40	55	61	73	72	96	134	238	213	279	223	211	1 781
Germany	9	83	45	66	77	81	76	110	122	166	142	177	186	224	1 564
United Kingdom	19	69	44	51	56	92	85	102	112	153	150	151	183	230	1 497
Japan	12	43	25	30	38	63	84	119	89	155	149	182	204	232	1 425
France	4	24	20	27	30	30	40	49	49	64	64	78	88	85	652
Switzerland	4	25	12	11	33	39	32	36	43	55	45	50	44	44	473
Netherlands	3	16	10	22	16	17	31	22	29	45	36	54	72	54	427
Israel	2	16	12	8	16	26	25	29	24	45	40	39	55	47	384
Belgium	2	17	4	9	10	9	14	29	17	25	28	50	44	49	307
Italy	2	1	8	6	9	12	13	21	20	22	27	31	20	44	236
Sweden	2	6	4	3	6	12	17	14	18	24	22	19	23	32	202
Australia	2	7	5	7	10	11	11	10	6	16	20	24	20	37	186
Chinese Taipei	1	3	5	5	7	6	7	2	5	13	13	24	26	31	148
South Korea	2	1		1	1		9	4	8	7	7	12	16	26	94
USSR[2]	2	1			1		1				3	1	3	12	24
All foreign countries	89	428	241	346	440	523	575	710	804	1 253	1 161	1 388	1 357	1 499	10 814

1. Note that patents are counted once for each individual inventor country of residence listed on a patent. As a result, some patents are counted more than once.
2. Data for the USSR include only those patents that list the Soviet Union as the "Country of residence of the applicant".

Source: USPTO TAF database, 11 February 1994.

Table 3. **Percentage of all US utility patents with foreign resident inventors also naming a US resident inventor, by country of residence of the foreign inventor**[1]

Selected countries

Country of residence of the foreign resident inventor	Year of patent grant														
	1980	1981	1982	1983	1984	1985	1986	1987	1988	1989	1990	1991	1992	1993	Total
Canada	1.4	5.9	3.9	5.2	4.9	5.3	5.3	5.8	8.3	10.7	10.1	12.0	10.4	10.1	7.9
Germany	0.2	1.3	0.8	1.2	1.2	1.2	1.1	1.4	1.6	1.9	1.8	2.2	2.5	3.1	1.6
United Kingdom	0.8	2.7	2.0	2.5	2.4	3.5	3.4	3.5	4.1	4.7	5.1	5.1	7.0	9.0	4.1
Japan	0.2	0.5	0.3	0.3	0.3	0.5	0.6	0.7	0.5	0.8	0.8	0.9	0.9	1.0	0.7
France	0.2	1.1	1.0	1.4	1.4	1.2	1.7	1.7	1.8	2.0	2.2	2.5	2.8	2.8	1.8
Switzerland	0.3	1.9	1.0	1.0	2.6	2.9	2.5	2.4	3.2	3.7	3.2	3.4	3.3	3.5	2.5
Netherlands	0.5	2.4	1.6	3.3	2.1	2.1	4.0	2.2	3.4	4.0	3.5	5.0	7.5	6.1	3.6
Israel	1.7	11.3	9.6	6.6	9.1	12.7	11.8	10.7	9.2	12.6	11.9	11.4	14.5	13.4	11.3
Belgium	0.8	5.7	1.7	3.8	3.8	3.4	5.3	8.6	5.0	6.0	7.4	12.4	10.8	11.2	6.8
Italy	0.2	0.1	1.0	0.9	1.1	1.3	1.3	1.7	1.8	1.6	2.1	2.5	1.5	3.3	1.6
Sweden	0.2	0.8	0.6	0.5	0.8	1.4	1.9	1.5	2.2	2.8	2.7	2.5	3.4	4.7	1.8
Australia	0.7	2.1	1.8	2.8	3.2	3.1	2.8	2.5	1.4	3.1	4.3	5.0	4.6	8.9	3.5
Chinese Taipei	1.5	3.6	5.3	7.2	6.7	3.4	3.3	0.6	1.1	2.2	1.8	2.6	2.6	2.6	2.4
South Korea	20.0	6.3	0.0	3.7	3.4	0.0	16.7	4.7	7.8	4.2	3.0	2.9	2.9	3.2	3.7
USSR[2]	0.4	0.3	0.0	0.0	0.5	0.0	0.8	0.0	0.0	0.0	1.7	0.6	4.1	15.6	0.9
All foreign countries	0.4	1.6	1.0	1.4	1.5	1.6	1.7	1.8	2.1	2.7	2.6	3.0	2.9	3.3	2.2

1. Note that patents are counted once for each individual inventor country of residence listed on a patent. As a result, some patents are counted more than once.
2. Data for the USSR include only those patents that list the Soviet Union as the "Country of residence of the applicant".

Source: USPTO TAF database, 11 February 1994.

have a relatively low propensity to co-invent with US resident inventors, with only 0.7 per cent of the patents with an inventor resident in Japan naming a US resident co-inventor. In general, patents naming inventors residing in Israel, Canada, Belgium, and the United Kingdom were more likely to include a US resident co-inventor, while patents naming inventors residing in Japan and the Soviet Union were less likely to include a US resident co-inventor. For the Soviet Union, however, it should be noted that the proportion of patents with Soviet Union resident inventors which also named a US resident co-inventor was 0 per cent in 1983 and rose to over 15 per cent in 1993, perhaps reflecting the increased willingness of US organisations to recruit Soviet technical assistance. Also notable is the 3.7 per cent share of patents with South Korean resident inventors that included a US co-inventor.

COLLABORATION BY TECHNOLOGY

US patents are classified according to the US Patent Classification System (USPCS). The USPCS consists of about 400 broad classes of technology that are divided into more than 130 000 subclasses. The USPCS evolved primarily as a classification scheme for use by the USPTO examiners; patent data organised by USPCS are thus difficult to use in conjunction with other socio-economic indicators based on more widely used systems, such as the Standard Industrial Classification system (SIC). For this reason, a concordance was created between the USPCS classifications and 41 unique product fields based on the 1972 SIC. This concordance is updated annually, as part of an inter-agency agreement between the USPTO and the National Science Foundation, to incorporate changes to the USPCS. Using this concordance, the US and foreign resident co-inventor patents were examined by product field; fractional counts, which divide patents equally between relevant SIC-based product fields, were used. More detail regarding fractional counts and the concordance may be found in OTAF (1984). This section presents the results of an analysis of international collaborative patents organised by SIC-based product fields and by USPCS.

International collaboration by SIC-based product fields

Patenting emphasis in a product field (*e.g.* SIC 20) for an identified patent category (*e.g.* all US and foreign co-inventor patents) has been determined by means of an activity index which equals the percentage of total patents in the product field for the identified patent category (112/7 824) divided by the percentage of patents in all product fields for the identified category (10 814/1 091 133). Thus, the activity index for SIC 20 for all US and foreign co-inventor patents equals 1.4 (Table 4).

Table 4. **Utility patent grants having a US resident inventor and a foreign resident inventor, 1980-93**

Patent counts as distributed by product fields based on the 1972 Standard Industrial Classification and miscellaneous categories

Product field title [SIC codes]	Total utility patents	Patent counts[1]				Activity index[2]			
		US-Japan co-inventor patents	US-foreign co-inventor patents	Ownership US org.	Type foreign org.	US-Japan co-inventor patents	US-foreign co-inventor patents	Ownership US org.	Type foreign org.
Food and kindred products [20]	7 824	11	112	68	40	1.1	1.4	1.3	2.1
Textile mill products [22]	6 992	0	69	46	20	0.0	1.0	1.0	1.2
Chemicals and allied products [28]	155 861	495	3 187	2 164	878	2.4	2.1	2.1	2.3
Chemicals, except drugs and medicines [281, 282, 284-289]	130 086	349	2 444	1 660	687	2.1	1.9	1.9	2.2
Basic industrial inorganic and organic chemistry [281, 286]	69 812	186	1 354	933	362	2.0	2.0	2.0	2.1
Industrial inorganic chemistry [281]	16 336	14	246	168	69	0.7	1.5	1.5	1.7
Industrial organic chemistry [286]	53 476	175	1 109	765	293	2.5	2.1	2.1	2.2
Plastics materials and synthetic resins [282]	17 532	49	318	209	104	2.1	1.8	1.8	2.4
Agricultural chemicals [287]	19 890	90	453	282	148	3.5	2.3	2.1	3.0
All other chemicals [284, 285, 289]	22 858	33	319	237	73	1.1	1.4	1.6	1.3
Soaps, detergents, cleaners, perfumes, cosmetics and toiletries [284]	6 882	4	89	72	15	0.4	1.3	1.6	0.9
Paints, varnishes, lacquers, enamels, and allied products [285]	6 502	21	112	78	31	2.5	1.7	1.8	1.9
Miscellaneous chemical products [289]	9 478	9	118	87	27	0.7	1.3	1.4	1.2
Drugs and medicines [283]	25 774	150	743	504	191	4.5	2.9	2.9	3.0
Petroleum and natural gas extraction and refining [13, 29]	11 745	7	178	152	21	0.5	1.5	1.9	0.7
Rubber and miscellaneous plastics products [30]	46 859	63	566	381	144	1.0	1.2	1.2	1.3
Stone, clay, glass and concrete products [32]	21 308	21	215	148	52	0.8	1.0	1.0	1.0

Table 4. **Utility patent grants having a US resident inventor and a foreign resident inventor, 1980-93** *(cont'd)*

Patent counts as distributed by product fields based on the 1972 Standard Industrial Classification and miscellaneous categories

Product field title [SIC codes]	Total utility patents	Patent counts[1]				Activity index[2]			
		US-Japan co-inventor patents	US-foreign co-inventor patents	Ownership US org.	Type foreign org.	US-Japan co-inventor patents	US-foreign co-inventor patents	Ownership US org.	Type foreign org.
Primary metals [33, 3462, 3463]	11 300	16	132	85	36	1.1	1.2	1.1	1.3
Primary ferrous products [331, 332, 3399, 3462]	5 150	10	61	39	16	1.5	1.2	1.1	1.3
Primary and secondary non-ferrous metals [333-336, 339 (except 3399), 3463]	6 154	10	71	46	20	1.2	1.2	1.1	1.3
Fabricated metal products [34 (except 3462, 3463, 348)]	83 549	34	463	304	87	0.3	0.6	0.5	0.4
Machinery, except electrical [35]	241 318	196	1 758	1 176	429	0.6	0.7	0.7	0.7
Engines and turbines [351]	10 915	8	49	23	18	0.6	0.5	0.3	0.7
Farm and garden machinery and equipment [352]	13 024	9	84	63	11	0.5	0.7	0.7	0.3
Construction, mining and material handling machinery and equipment [353]	26 573	12	185	125	42	0.3	0.7	0.7	0.6
Metal-working machinery and equipment [354]	25 698	10	119	74	36	0.3	0.5	0.4	0.6
Office computing and accounting machines [357]	41 498	104	460	339	102	1.9	1.1	1.2	1.0
Other machinery, except electrical [355, 356, 358, 359]	123 611	58	861	551	221	0.4	0.7	0.7	0.7
Special industry machinery, except metal-working [355]	45 914	23	333	198	102	0.4	0.7	0.6	0.9
General industrial machinery and equipment [356]	55 091	25	351	231	82	0.3	0.6	0.6	0.6
Refrigeration and service industry machinery [358]	15 947	7	139	98	26	0.3	0.9	0.9	0.7
Miscellaneous machinery, except electrical [359]	6 659	4	37	24	10	0.5	0.6	0.5	0.6

Table 4. **Utility patent grants having a US resident inventor and a foreign resident inventor, 1980-93** *(cont'd)*

Patent counts as distributed by product fields based on the 1972 Standard Industrial Classification and miscellaneous categories

Product field title [SIC codes]	Total utility patents	Patent counts[1]				Activity index[2]			
		US-Japan co-inventor patents	US-foreign co-inventor patents	Ownership US org.	Type foreign org.	US-Japan co-inventor patents	US-foreign co-inventor patents	Ownership US org.	Type foreign org.
Electrical and electronic machinery, equipment and supplies [36, 3825]	222 015	335	2 007	1 392	491	1.2	0.9	0.9	0.9
Electrical equipment, except communications equipment [361-364, 369, 3825]	77 080	90	668	439	167	0.9	0.9	0.9	0.9
Electrical transmission and distribution equipment [361, 3825]	26 532	48	282	202	70	1.4	1.1	1.1	1.1
Electrical industrial apparatus [362]	23 567	20	164	95	41	0.6	0.7	0.6	0.7
Other electrical machinery, equipment and supplies [363, 364, 369]	26 981	25	222	141	56	0.7	0.8	0.8	0.8
Household appliances [363]	7 817	6	49	34	8	0.6	0.6	0.7	0.4
Electrical lighting and wiring equipment [364]	9 293	5	53	32	13	0.4	0.6	0.5	0.6
Miscellaneous electrical machinery, equipment and supplies [369]	9 876	16	120	75	35	1.2	1.2	1.1	1.4
Communications equipment and electronic components [365-367]	144 937	247	1 339	954	324	1.3	0.9	1.0	0.9
Radio and television receiving equipment except communication types [365]	12 787	20	81	54	24	1.2	0.6	0.6	0.8
Electronic components and accessories and communications equipment [366-367]	132 149	229	1 258	899	300	1.3	1.0	1.0	0.9
Transportation equipment [37, 348]	53 911	37	228	113	79	0.5	0.4	0.3	0.6
Motor vehicles and other transportation equipment, except aircraft [348, 371, 373-376, 379]	41 679	25	175	87	58	0.5	0.4	0.3	0.6
Motor vehicles and other motor vehicle equipment [371]	23 744	19	90	43	35	0.6	0.4	0.3	0.6
Guided missiles and space vehicles and parts [376]	722	0	7	7	0	0.0	1.0	1.5	0.0

Table 4. **Utility patent grants having a US resident inventor and a foreign resident inventor, 1980-93** *(cont'd)*

Patent counts as distributed by product fields based on the 1972 Standard Industrial Classification and miscellaneous categories

Product field title [SIC codes]	Total utility patents	Patent counts[1]				Activity index[2]			
		US-Japan co-inventor patents	US-foreign co-inventor patents	Ownership US org.	Type foreign org.	US-Japan co-inventor patents	US-foreign co-inventor patents	Ownership US org.	Type foreign org.
Other transportation equipment [373-375, 379 (except 3795)]	11 766	8	58	28	16	0.5	0.5	0.4	0.6
Ship and boat building and repairing [373]	3 290	3	20	8	6	0.7	0.6	0.4	0.7
Railroad equipment [374]	2 377	1	12	8	3	0.3	0.5	0.5	0.5
Motorcycles, bicycles, and parts [375]	2 138	0	9	5	2	0.0	0.4	0.4	0.4
Miscellaneous transportation equipment [379 (except 3795)]	3 957	3	17	7	5	0.6	0.4	0.3	0.5
Ordinance except missiles [348, 3795]	5 447	0	20	9	7	0.0	0.4	0.2	0.5
Aircraft and parts [372]	12 234	10	53	26	21	0.6	0.4	0.3	0.7
Professional and scientific instruments [38 (except 3825)]	141 094	171	1 377	941	278	0.9	1.0	1.0	0.8
All other SICs [99]	87 363	51	523	299	125	0.4	0.6	0.5	0.6
All industries (total)[3]	1 091 133	1 425	10 814	7 268	2 680	1.0	1.0	1.0	1.0

1. Patent counts are determined by means of a concordance between the US Classification System (USPCS) and the SIC-based product fields. A fractional counting procedure has been used wherein patents receiving original classification in a USPCS sub-class are evenly divided between the SIC-based product fields assigned to the sub-class. More detail on the concordance, may be found in OTAF (1984). Ownership type is determined by the first-named assignee at the time of grant (US or foreign organisation, including both government and non-government organisations).
2. Activity index figures indicate the degree to which patenting activity is emphasized for a SIC-based product field. It is calculated by dividing the patent share in a product field by the patent share in all product fields (all industries). Indexes greater than 1 indicate a greater than average patenting emphasis.
3. Owing to rounding, row and column totals may not equal the sum of the row and column elements.

Source: USPTO TAF database, 11 January 1994.

During the 1980-93 period, the technologies most emphasized in patents by US and foreign resident co-inventors were related to the unique product fields of industrial organic chemistry (SIC 286), plastics materials and synthetic resins (SIC 282), agricultural chemicals (SIC 287), and drugs and medicines (SIC 283), all included within the broad product field chemicals and allied products (SIC 28) (see Table 5).

In general, for the US and foreign co-inventor patents for which ownership was assigned to a US organisation at the time of granting, patenting emphasized the same technologies. However, US organisation-owned patents also emphasized technologies relating to petroleum and natural gas extraction and refining (SIC 13, 29), an area where the United States has traditionally been very active (National Science Foundation, 1993, Appendix Table 6-13). Thus, US organisations appear to seek relatively more foreign technical expertise in technologies relating to SIC 286, 282, 287, 283, 13, and 29).

The subset of collaborative patents for which ownership was assigned to a foreign organisation emphasized technologies related to the same product fields as those identified for patents assigned to US organisations, with the omission of petroleum and natural gas extraction and refining (SIC 13, 29). Food and kindred products (SIC 20) and paints, varnishes, lacquers, enamels, and allied products (SIC 285), were also emphasized, which suggests that foreign organisations may seek relatively more US technical expertise in these areas as well.

For the patents that included US and Japanese resident co-inventors, patenting emphasized the technologies related to the same product fields as patents for US and foreign resident co-inventors, with the addition of paints, varnishes, lacquers, enamels, and allied products (SIC 285) and office computing and accounting machines (SIC 357).

International collaboration by USPCS technology

A similar index was used to determine patenting emphasis by class within the USPCS (Table 6). In general, the findings were consistent with the SIC-based findings. For all patents having US and foreign co-inventors, chemical technologies were the most emphasized, along with induced nuclear reaction technology. For patents with US and foreign resident co-inventors for which ownership was assigned to a US organisation (either government or non-government), based on the first-named assignee), the most highly emphasized technologies generally reflected those emphasized in all patents having US and foreign resident co-inventors. In addition, technologies pertaining to wells (for the extraction of fluids from the earth) and to organisation of information processing systems were identified.

Table 5. **Utility patent grants having a US resident inventor and a foreign resident inventor, 1980-93**

Patent counts[1] as distributed by product fields based on the 1972 Standard Industrial Classification and year of grant

	Year of patent grant														
	1980	1981	1982	1983	1984	1985	1986	1987	1988	1989	1990	1991	1992	1993	Total
Food and kindred products [20]	2	4		1	5	5	6	4	6	18	14	20	16	11	112
Textile mill products [22]		3	1	8	4	4	7	4	3	6	6	8	9	8	69
Chemicals and allied products [28]	30	150	93	121	135	152	164	222	212	304	346	364	419	477	3 187
Chemicals, except drugs and medicines [281, 282, 284-289]	26	124	67	92	116	116	132	165	166	236	256	286	319	346	2 444
Basic industrial inorganic and organic chemistry [281, 286]	18	98	41	54	75	60	78	86	90	130	147	143	156	182	1 354
Industrial inorganic chemistry [281]	4	15	12	11	17	13	20	17	14	26	22	23	24	31	246
Industrial organic chemistry [286]	15	83	29	43	58	47	58	69	76	104	125	120	132	152	1 109
Plastics materials and synthetic resins [282]	1	8	10	11	15	17	14	19	20	34	29	35	53	54	318
Agricultural chemicals [287]	5	12	13	20	16	21	19	38	27	51	53	54	59	67	453
All other chemicals [284, 285, 289]	3	7	3	7	10	19	21	22	28	22	28	55	52	43	319
Soaps, detergents, cleaners, perfumes, cosmetics and toiletries [284]		3	1	2	4	5	3	10	9	6	8	15	11	14	89
Paints, varnishes, lacquers, enamels, and allied products [285]	1	3	2	4	5	10	7	7	10	12	7	21	18	9	112
Miscellaneous chemical products [289]	2	1	1	1	2	4	12	6	10	4	13	19	24	20	118
Drugs and medicines [283]	4	26	27	29	20	36	32	58	47	68	90	78	100	131	743
Petroleum and natural gas extraction and refining [13, 29]	1	3	4	12	7	10	9	12	9	27	15	28	20	24	178
Rubber and miscellaneous plastics products [30]	6	16	7	13	25	27	27	31	41	66	63	80	82	80	566
Stone, clay, glass and concrete products [32]	1	9	1	7	10	9	9	14	17	25	23	38	30	24	215
Primary metals [33, 3462, 3463]	2	8	1	5	6	9	8	6	14	20	11	12	14	20	132
Primary ferrous products [331, 332, 3399, 3462]	2	5	1	4	3	4	2	1	7	7	4	4	8	9	61

Table 5. **Utility patent grants having a US resident inventor and a foreign resident inventor, 1980-93** *(cont'd)*

Patent counts[1] as distributed by product fields based on the 1972 Standard Industrial Classification and year of grant

	Year of patent grant														
	1980	1981	1982	1983	1984	1985	1986	1987	1988	1989	1990	1991	1992	1993	Total
Primary and secondary non-ferrous metals [333-336, 339 (except 3399), 3463]		2	1	1	3	5	6	5	7	13	7	7	6	11	71
Fabricated metal products [34 (except 3462, 3463, 348)]	2	21	6	9	19	25	24	32	41	77	60	61	32	57	463
Machinery, except electrical [35]	14	58	38	59	71	77	87	104	142	204	192	256	213	243	1 758
Engines and turbines [351]	1	1		3	1	4	3	4	3	5	6	5	7	6	49
Farm and garden machinery and equipment [352]	1	4	1	1	2	6	4	7	7	10	14	13	4	9	84
Construction, mining and material handling machinery and equipment [353]	3	7	1	7	6	10	9	17	18	18	18	25	22	26	185
Metal-working machinery and equipment (354)	1	4	3	8	5	3	7	4	11	22	9	18	13	12	119
Office computing and accounting machines [357]	1	20	11	11	18	14	22	21	33	56	47	62	56	88	460
Other machinery, except electrical [355, 356, 358, 359]	7	22	21	30	40	41	43	51	71	93	98	133	110	103	861
Special industry machinery, except metal-working [355]	5	11	9	16	19	19	18	20	31	40	33	38	35	37	333
General industrial machinery and equipment [356]	1	7	8	9	13	14	17	23	25	35	45	61	51	43	351
Refrigeration and service industry machinery [358]	1	3	2	4	6	6	6	6	12	13	17	28	17	18	139
Miscellaneous machinery, except electrical [359]		1	1		2	2	2	2	3	4	2	6	7	5	37
Electrical and electronic machinery, equipment and supplies [36, 3825]	13	64	39	38	68	98	103	144	157	255	231	261	267	269	2 007
Electrical equipment, except communications equipment [361-364, 369, 3825]	6	19	15	13	26	37	35	54	57	75	83	100	75	74	668
Electrical transmission and distribution equipment [361, 3825]	1	7	6	2	11	13	16	23	28	28	41	43	30	34	282
Electrical industrial apparatus [362]	4	5	2	3	10	8	10	11	14	24	20	26	16	14	164

Table 5. **Utility patent grants having a US resident inventor and a foreign resident inventor, 1980-93** *(cont'd)*

Patent counts[1] as distributed by product fields based on the 1972 Standard Industrial Classification and year of grant

	Year of patent grant														
	1980	1981	1982	1983	1984	1985	1986	1987	1988	1989	1990	1991	1992	1993	Total
Other electrical machinery, equipment and supplies [363, 364, 369]	1	7	8	8	6	17	9	20	15	23	22	31	29	27	222
Household appliances [363]		2	2	1	1	4	2	5	6	7	4	6	5	3	49
Electrical lighting and wiring equipment [364]		4	1	2	1	3	2	4	3	3	10	9	8	6	53
Miscellaneous electrical machinery, equipment and supplies [369]	1	2	5	6	4	10	5	12	6	13	9	16	17	18	120
Communications equipment and electronic components [365-367]	8	45	25	25	42	61	68	90	100	180	149	160	191	195	1 339
Radio and television receiving equipment except communication types [365]	1	4		1	2	7	6	7	5	11	9	5	10	15	81
Electronic components and accessories and communications ent [366-367]	7	41	25	25	40	55	62	83	95	170	140	155	181	180	1 258
Transportation equipment [37, 348]	2	7	3	4	12	14	15	21	12	24	30	22	31	32	228
Motor vehicles and other transportation equipment, except [348, 371, 373-376, 379]	1	6	2	3	8	11	11	15	10	18	23	16	24	27	175
Motor vehicles and other motor vehicle equipment [371]	1	1	1	2	4	6	6	10	4	10	10	7	15	14	90
Guided missiles and space vehicles and parts [376]						1	1	1		2	1				7
Other transportation equipment [373-375, 379 (except 3795)]		5		1	3	4	4	4	4	6	8	6	7	8	58
Ship and boat building and repairing [373]		3		1	2	2	2		4	2		2	2	1	20
Railroad equipment [374]		1				1					5	2	1	2	12

Table 5. **Utility patent grants having a US resident inventor and a foreign resident inventor, 1980-93** *(cont'd)*

Patent counts[1] as distributed by product fields based on the 1972 Standard Industrial Classification and year of grant

	Year of patent grant														
	1980	1981	1982	1983	1984	1985	1986	1987	1988	1989	1990	1991	1992	1993	Total
Motorcycles, bicycles, and parts [375]						1				1	1	1	2	3	9
Miscellaneous transportation equipment [379 (except 3795)]		1				1	2	3		2	2	1	2	2	17
Ordinance except missiles [348, 3795]			1		2		1		2	1	5	2	2	4	20
Aircraft and parts [372]	1	1	1	2	4	3	3	6	2	5	6	6	8	5	53
Professional and scientific instruments [38 (except 3825)]	11	51	36	49	50	60	73	82	127	179	131	174	169	185	1 377
All other SIC's [99]	5	35	13	21	28	33	47	35	23	48	42	67	56	71	523
All industries[2]	89	428	241	346	440	523	575	710	804	1 253	1 161	1 388	1 357	1 499	10 814

1. Patent counts are determined by means of a concordance between the US Classification System (USPCS) and the SIC-based product fields. A fractional counting procedure has been used wherein patents receiving original classification in a USPCS sub-class are evenly divided among the SIC-based product fields assigned to the sub-class. More detail on the concordance, may be found in OTAF (1984).
2. Owing to rounding, row and column totals may not equal the sum of the row and column elements.

Source: USPTO TAF database, 11 January 1994.

Table 6. **Classes of technology most emphasized by utility patent grants to US resident and foreign resident co-inventors 1980-93**

All utility patents having US and foreign resident co-inventors

Patent class title	Activity index[1] Class	1980-93
Chemistry: natural resins or derivatives; peptides or proteins; lignins or reaction products thereof	530	4.44
Chemistry: molecular biology and microbiology	435	3.46
Drug, bio-affecting and body treating compositions	514	2.56
Organic compounds – part of the class 532-570 series	548	2.53
Drug, bio-affecting and body treating compositions	424	2.44
Chemistry: analytical and immunological testing	436	2.25
Induced nuclear reactions: processes, systems, and elements	376	2.15
Synthetic resins or natural rubbers – part of the class 520 series	525	2.11
Chemistry of hydrocarbon compounds	585	2.03
Organic compounds – part of the class 532-570 series	549	2.00

US organisation (government and non-government) owned utility patents having US and foreign resident co-inventors

Patent class title	Activity index[1] Class	1980-93
Chemistry: natural resins or derivatives; peptides or proteins; lignins or reaction products thereof	530	4.95
Chemistry: molecular biology and microbiology	435	3.83
Induced nuclear reactions: processes, systems, and elements	376	2.91
Chemistry of hydrocarbon compounds	585	2.63
Drug, bio-affecting and body treating compositions	424	2.52
Drug, bio-affecting and body treating compositions	514	2.35
Chemistry: analytical and immunological testing	436	2.35
Organic compounds – part of the class 532-570 series	548	2.28
Wells	166	2.26
Information processing system organisation	395	2.22

Foreign organisation (government and non-government) owned utility patents having US and foreign resident co-inventors

Patent class title	Activity index[1] Class	1980-93
Chemistry: natural resins or derivatives; peptides or proteins; lignins or reaction products thereof	530	3.61
Organic compounds – part of the class 532-570 series	548	3.55
Drug, bio-affecting and body treating compositions	514	3.40
Synthetic resins or natural rubbers – part of the class 520 series	525	2.89
Synthetic resins or natural rubbers – part of the class 520 series	528	2.84
Metal founding	164	2.63
Plant protecting and regulating compositions	504	2.59
Synthetic resins or natural rubbers – part of the class 520 series	524	2.52
Chemistry: molecular biology and microbiology	435	2.48
Organic compounds – part of the class 532-570 series	544	2.37

1. For a category, *e.g.* all US and foreign resident co-inventor patents, the activity index for a class is the percentage of category patents in that class divided by the percentage of all patents in that class. Indexes greater than 1 indicate a greater than expected emphasis on patenting. Patent classes are limited to those in which 2 800 patents were granted during 1980-93.

Source: TAF database, 11 February 1994.

OWNERSHIP CATEGORY

Invention by US and foreign resident co-inventors is primarily an activity of corporate or government organisations, as opposed to individuals. Two-thirds of the patents naming US and foreign co-inventors were assigned to US corporate or government organisations and one-quarter were assigned to foreign corporate or government organisations (Tables 7 and 8). The remaining 8 per cent of patents with US and foreign resident co-inventors were assigned to an individual (or unassigned). By comparison, only 25 per cent of all patents including a US resident inventor during the 1980-93 period were assigned to an individual (or unassigned) and 11 per cent of all patents including a foreign resident inventor were assigned to an individual (or unassigned) (these latter two data elements are not displayed in the attached tables). Ownership category shares as a percentage of the total have fluctuated somewhat and have not displayed either a consistent upward or downward trend during the 1980-93 period.

ANALYSIS OF PATENT ASSIGNMENT

Patents with international co-inventorship may have several sources involving global research networks. For example, they mights arise from: *i)* research activity between parent firms and foreign subsidiaries; *ii)* research activity between unrelated firms, governments and research institutes; or *iii)* research activity stemming from intra-corporate arrangements in which a firm's employees reside in several countries. Factors such as corporate policies that involve compensation or award incentives may also influence the formation of teams of inventors (Aspden, 1983).

Issues concerning the use of patent ownership data

Analysis of US patent ownership is complicated by factors that can be traced back to the process that produces the ownership information. Researchers using US patent ownership data must be knowledgeable about the source of these data and about potential problems with their use. This warrants a brief discussion of the source of ownership data and potential problems.

Before an approved patent application is issued, the applicant, or the applicant's legal agent or representative, is asked to provide the assignee name(s) that will appear on the patent. Assignee names can be individuals, governments, corporations or other entities. The assignee information that is provided by the applicant is recorded on the patent, except when the applicant provides several names for the same entity, in which case, only the first name will appear on the patent. For example, Jones and Jones, Inc., doing business as the Jones Brothers will appear on the patent as Jones and Jones, Inc.[3] If the applicant does not provide an assignee name to appear on the patent, the patent will be issued as an unassigned patent, even if ownership rights have been assigned during the processing of the application.

Table 7. **Number of US utility patent grants with a US resident inventor and a foreign resident inventor, distributed by ownership category at grant and year of grant**

Ownership category[1]	Year of patent grant														
	1980	1981	1982	1983	1984	1985	1986	1987	1988	1989	1990	1991	1992	1993	Total
US corporations	47	267	155	225	286	334	346	432	531	838	772	924	877	996	7 030
US government	2	10	6	10	19	23	21	24	25	19	16	12	23	28	238
US individuals	8	34	13	21	21	20	38	31	28	67	65	78	60	55	539
Foreign corporations	29	114	63	89	95	121	142	198	193	288	259	329	350	377	2 647
Foreign government			1		1	4	3		3	3	4	2	9	3	33
Foreign individuals	3	3	3	1	18	21	25	25	24	38	45	43	38	40	327
All ownership	89	428	241	346	440	523	575	710	804	1 253	1 161	1 388	1 357	1 499	10 814

1. Ownership category is based on the first-named assignee, as displayed on the patent. The "US corporations" category includes all non-federal government organisations assigned patent ownership at the time of grant. The "foreign corporations" category includes all non-government organisations assigned patent ownership at the time of grant. These categories consist predominantly but not exclusively of corporations (*i.e.* small businesses, universities, non-profit organisations, etc., are also included in this category).

Source: USPTO TAF database, 11 March 1994.

Table 8. **US utility patent grants with a US resident inventor and a foreign resident inventor**

Share of total "all ownership" patents (Table 4) distributed by ownership category at grant and the year of grant

Percent

Ownership category[1]	Year of patent grant														
	1980	1981	1982	1983	1984	1985	1986	1987	1988	1989	1990	1991	1992	1993	Total
US corporations	53	62	64	65	65	64	60	61	66	67	66	67	65	66	65
US government	2	2	2	3	4	4	4	3	3	2	1	1	2	2	2
US individuals	9	8	5	6	5	4	7	4	3	5	6	6	4	4	5
Foreign corporations	33	27	26	26	22	23	25	28	24	23	22	24	26	25	24
Foreign government	0	0	0	0	0	1	1	0	0	0	0	0	1	0	0
Foreign individuals	3	1	1	0	4	4	4	4	3	3	4	3	3	3	3
All ownership	100	100	100	100	100	100	100	100	100	100	100	100	100	100	100

1. Ownership category is based on the first-named assignee, as displayed on the patent. The "US corporations" category includes all non-federal government organisations assigned patent ownership at the time of grant. The "foreign corporations" category includes all non-government organisations assigned patent ownership at the time of grant. These categories consist predominantly but not exclusively of corporations (*i.e.* small businesses, universities, non-profit organisations, etc. are also included in this category).

Source: USPTO TAF database, 11 March 1994.

Patents may be owned by multiple entities, and all joint owners, no matter how small their interest, have the same rights. A joint owner can prevent others from making, using, or selling the invention; he or she can also sell the interest or any part of it and grant licences to others, independently of the other owner(s). Thus, it is extremely prudent for joint owners of a patent to have an agreement governing these rights.

Clearly, it is the responsibility of the applicant to identify the assignee at the time of issue, and he or she has a considerable degree of latitude with respect to who is named and how that name appears. This presents at least two methodological problems for the researcher. First, the assignee name on the patent grant can contain typographical errors and differences in form. As a result, a single assignee may appear under several names. There are, for example, 70 variations of the name Leland Stanford University in the APS database (The Board of Regents of...; Stanford University; Regents of..., etc.). Fortunately, the TAF database incorporates many corrections for typographical errors and reduces many formal variations.

The second, and more difficult, methodological issue is that parent, subsidiary, or divisional units of a firm are sometimes used interchangeably. This is particularly problematic for research on the globalisation of technology, which involves study of patent activity by foreign subsidiaries. Inconsistent use of firm names may result from corporate policy, prior legal agreements, or decision by the legal representative.[4] Some firms hire several legal representatives, who may use different parent and subsidiary names. Whatever the the reason, patent records furnish researchers with compelling evidence for caution in using patent data to analyse global organisational networks of technological development. Two examples of US patent activity by a British firm and a Japanese firm illustrate these issues.

US patent 5 281 578 was assigned to Johnson Matthey Public Limited Company in 1994.

US PAT NO:	5 281 578
DATE FILED:	19 May 1992
DATE ISSUED:	25 January 1994
INVENTOR:	Fontaine C. Bradley, Byfield, MA Danielle T. Frost, Philadelphia, PA **Christen M. Giandomenico, West Chester, PA**
ASSIGNEE:	Johnson Matthey Public Limited Company, London, England (foreign corp.)

Based on assignment information alone, it would appear that this patent represents US patent activity by a foreign firm. When the inventors listed on the patent are identified and considered, however, it appears that it incorporated US inventive resources that resulted in a patent to a foreign firm. Again, this may not precisely describe the situation. Johnson Matthey Public Limited Company operates a subsidiary in the United States called Johnson Matthey, Inc. A review of additional patent records provides additional evidence to suggest that patent 5 281 578 resulted from work performed by the subsidiary. Patent 5 244 919, for example, was assigned to the subsidiary in Valley Forge, Pennsylvania; it also lists one of the inventors (Christen M. Giandomenico) listed on patent 5 281 578.

US PAT NO:	5 244 919
DATE FILED:	1 July 1991
DATE ISSUED:	14 September 1993
INVENTOR:	Michael J. Abrams, Glenmore, PA
	Christen M. Giandomenico, Exton, PA
	Barry A. Murrer, Reading, United Kingdom
	Jean F. Vollano, Exton, PA
ASSIGNEE:	Johnson Matthey , Inc., Valley Forge, PA (US corp.)

Additional patent records indicate that these two entities have undertaken collaborative research since 1990 (although both entities have been in existence much longer). There is no clear reason why the parent firm was named on patent 5 281 578; this is particularly curious since all of the inventors were US residents and other patents containing the names of these inventors were assigned to the US subsidiary.

A slight variation of this methodological issue is presented by the record of US patents assigned to Eisai Co., Ltd., a Japanese firm that was granted patent 5 273 985 in 1993.

US PAT NO:	5 273 985
DATE FILED:	4 June 1991
DATE ISSUED:	28 December 1993
INVENTOR:	Kazuo Okano, Ibaraki, Japan
	Osamu Asano, Ibaraki, Japan
	Naoyuki Shimomura, Ibaraki, Japan
	Tetsuya Kawahara, Ibaraki, Japan
	Shinya Abe, Ibaraki, Japan
	Shuhei Miyazawa, Ibaraki, Japan
	Mitsuaki Miyamoto, Waltham, MA
	Hiroyuki Yoshimura, Ibaraki, Japan
	Koukichi Harada, Ibaraki, Japan
	Junsaku Nagaoka, Ibaraki, Japan
	Tsutomu Kawata, Waltham, MA
	Tsutomu Yoshimura, Ibaraki, Japan
	Hiromasa Suzuki, Ibaraki, Japan
	Shigeru Souda, Ibaraki, Japan
	Yoshimasa Machida, Ibaraki, Japan
	Kouichi Katayama, Ibaraki, Japan
	Isao Yamatsu, Ibaraki, Japan
ASSIGNEE:	Eisai Co., Ltd., Tokyo, Japan (foreign corp.)

This patent was a multinational, collaborative effort between US and Japanese resident inventors; two of the inventors, Mitsuaki Miyamoto and Tsutomu Kawata, resided in Waltham, Massachusetts. It is conceivable that the patent represents a research effort involving technical expertise from both US and Japanese sources. However, additional evidence, albeit circumstantial, suggests otherwise. First, the names of the two US resident inventors also appear as Japan-resident inventors on other patents assigned to Eisai Co., Ltd. Mitsuaki Miyamoto, for example, is listed as a resident of Ibaraki, Japan on patent 5 221 671:

US PAT NO:	5 221 671
DATE FILED:	17 October 1991
DATE ISSUED:	22 June 1993
ASSIGNEE:	Eisai Co., Ltd., Tokyo, Japan (foreign corp.)
INVENTOR:	Kazuo Okano, Ibaraki, Japan
	Shuhei Miyazawa, Ibaraki, Japan
	Richard S.J. Clark, Ibaraki, Japan
	Shinya Abe, Ibaraki, Japan
	Tetsuya Kawahara, Ibaraki, Japan
	Naoyuki Shimomura, Ibaraki, Japan
	Osamu Asano, Ibaraki, Japan
	Hiroyuki Yoshimura, Ibaraki, Japan
	Mitsuaki Miyamoto, Ibaraki, Japan
	Yoshimori Sakuma, Ibaraki, Japan
	Kenzo Muramoto, Ibaraki, Japan
	Hiroshi Obaishi, Ibaraki, Japan
	Koukichi Harada, Ibaraki, Japan
	Hajime Tsunoda, Ibaraki, Japan
	Satoshi Katayama, Ibaraki, Japan
	Kouji Yamada, Ibaraki, Japan
	Shigeru Souda, Ibaraki, Japan
	Yoshimasa Machida, Ibaraki, Japan
	Kouichi Katayama, Ibaraki, Japan
	Isao Yamatsu, Ibaraki, Japan
ASSIGNEE:	Eisai Co., Ltd., Tokyo, Japan (foreign corp.)

It is possible that patent 5 273 985 was an effort involving the Eisai Research Institute of Boston, Inc., a subsidiary of Eisai Co., Ltd., which is located in Andover, Massachusetts. Unfortunately, the patent record does not reflect this; in fact, the US patent file does not contain patents assigned to the Eisai Research Institute of Boston, Inc.

These two examples illustrate potential problems for the analysis of ownership data. In both, foreign subsidiaries may have been involved in the patent activity, even though the patent record does not show this. In the first example, a patent assigned to Johnson Matthey Public Limited Company may have resulted from inventive activity sponsored by a US subsidiary using US inventors. In the second, what at first appears to be a collaborative effort by multinational inventors may in fact be a relocation of parent firm resources to a foreign subsidiary. Both examples provide evidence of international collaboration, but the pattern of patenting activity is not clear.

Investigative research involving a smaller sample of patents may help disclose more precisely how international patent activity took place. Patent files or the assignees and inventors could be consulted for information concerning the national origin of the collaborative effort. The amount of bias present as a result of the variable procedures and practices of attorneys and firms when specifying the assignee is unknown; however, evidence in the patent records suggests that this may be a significant problem.

One way to reduce the bias in assignee data might be to identify the parent firm(s) of each assignee. While this will help to reduce the bias caused by variously assigning patents to the parent firm or a subsidiary, it introduces a new problem. The dynamism of multinational firms results from frequent acquisitions and mergers; therefore, a 1994 reference might identify the parent of another entity that received a patent in 1990; however, the latter could have been a subsidiary of another firm when the patent was granted. As a result, this approach is less precise if historical assignment data are related to contemporary reference materials. For this reason, the present research was restricted to organisations that were actively patenting between 1989 and 1993. Patents assigned to subsidiaries were not combined with patents assigned to parent firms, since the aim was to obtain at least a minimum measurement of the volume of international collaboration by foreign subsidiaries in the United States. Furthermore, the analysis was restricted to patents owned by organisations. Preliminary analysis of ownership information showed that only about 5 per cent of the patents were assigned to individuals who were US residents and another 3 per cent to individuals who were foreign residents. Unassigned patents have not been treated.

Patenting by organisations

Over three-fourths of all US utility patents are assigned to corporations and other private organisations. The others are assigned to independent inventors or government agencies. Patents covered in this study of international collaboration were even more likely (89 per cent) to be assigned to corporations and private organisations. The analysis of patenting by organisation focused on government, corporate, and research organisations that were significantly involved in international collaboration; these were defined as organisations that had been assigned five or more collaborative patents since 1980 and were still actively patenting between 1989 and 1993. As a result, 360 organisations were identified; they were assigned 4 036 collaborative patents between 1989 and 1993, or about three-fourths of all collaborative patents assigned during that period.[5] Next, each assignee was manually classified into one of six organisational categories, primarily on the basis of information in US and international corporate directories and TAF reports.[6] The categories are:

- a US subsidiary of a foreign-owned firm;
- a foreign subsidiary of a US-owned firm;
- a foreign-owned firm;
- a US firm;
- a US college or university;
- an agency of the US government.

Analysis showed that the majority of international collaborative patents were assigned to US corporations (55 per cent), followed by foreign corporations (21 per cent), and US subsidiaries of foreign corporations (12 per cent) (see Table 9). However, the role of foreign subsidiaries in collaborative patent activity may be much greater than these data indicate, since the data may reflect a tendency on the part of legal representatives to record parent firm names rather than subsidiary names.

US colleges and universities were assigned a respectable 8 per cent of the collaborative patents. This is consistent with prior research indicating that US academic institutions provide non-US residents with valuable access to research opportunities. The relatively high incidence of collaboration in the academic community is also reflected in Figure 1, which compares collaborative patent activity to total patent activity for assignees in each of the six organisational categories.[7] An index of 1 would result, for example, if 10 per cent of the collaborative patents were assigned to the entities in a particular organisational group that also had 10 per

Table 9. **Number and percentage of patents with US and foreign co-inventors assigned to organisations within each organisational category, utility patents, 1989-93**

Organisational category	Patents with US and foreign inventors	
	Number	Per cent
US corporation	2 253	55.4
Foreign corporation[1]	863	21.4
US subsidiary of foreign corporation	474	11.7
US university	309	7.7
US government	99	2.4
Foreign subsidiary of US corporation	56	1.4
Total	4 036	100.0

1. This category includes foreign government and academic institutions.
Source: APS database, 11 February 1994

◆ Figure 1. ***Index of collaboration versus percentage of collaborative patents***

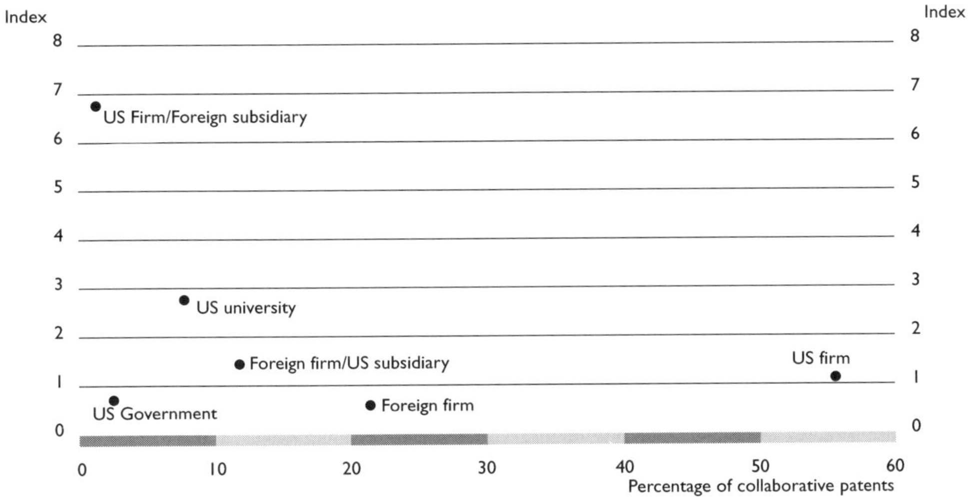

Source: Author.

cent of the total patent activity (collaborative and non-collaborative). Index values greater or less than 1 indicate proportionally greater or fewer numbers of collaborative patents.

The data in Figure 1 show that US universities have a relatively small percentage of collaborative patents (8 per cent), but they received nearly three times the expected number of collaborative patents (index = 2.8). While foreign subsidiaries of US companies received the smallest percentage of collaborative patents (1 per cent), they received nearly seven times the expected number of collaborative patents (index = 6.8).

Index values were also used to identify the countries with corporate entities that were assigned collaborative patents. The national identity of 99 non-US firms and parents of 39 US subsidiaries of non-US firms were manually determined from corporate directories. Index values were then calculated for all firms associated with each country. Data presented in Table 10 reaffirm earlier findings concerning the national residence of foreign inventors. A very high percentage of the collaborative patents were assigned to Japanese firms (9 per cent), yet collaboration with US resident inventors was less likely to occur in all patents assigned to Japanese firms (index = .4). Conversely, Israeli and Italian firms were assigned a very small share of collaborative patents (.3 per cent and .5 per cent respectively), yet collaboration with US resident inventors was more likely to occur in patents assigned to firms in these nations than in other nations (index = 5.5 and 3.2, respectively).

Table 10. **Foreign organisations and US subsidiaries of foreign organisations that were assigned US utility patents with US and foreign co-inventors, 1989-93**

	Foreign organisations			US subsidiary of foreign organisation			Total		
	Number of organisations	Collaborative patents (%)	Index	Number of organisations	Collaborative patents (%)	Index	Number of organisations	Collaborative patents (%)	Index
Japan	36	0.1	0.3	3	0.0	6.0	39	0.1	0.4
Germany	15	0.1	0.8	7	0.0	2.5	22	0.1	1.1
United Kingdom	9	0.0	1.8	4	0.0	2.5	13	0.0	2.0
Canada	13	0.0	2.5				13	0.0	2.5
France	7	0.0	1.9	7	0.0	1.1	14	0.0	1.4
Switzerland	5	0.0	1.3	5	0.0	1.4	10	0.0	1.4
Italy	3	0.0	3.1	1	0.0	3.3	4	0.0	3.2
Israel	2	0.0	5.5				2	0.0	5.5
Netherlands	2	0.0	0.7	10	0.0	1.0	12	0.0	0.9
Others	7	0.0	2.0	2	0.0	2.7	9	0.0	2.2
Total	99	0.2	0.6	39	0.1	1.5	138	0.3	0.8

1. Index values relate the number of US patents with US and foreign co-inventors assigned to organisations within each country and the total number of US patents assigned to those organisations.

Source: USPTO APS database, 11 February 1994.

IDENTIFICATION OF COLLABORATIVE PATENTING ACTIVITY – FURTHER DISCUSSION

Examining patent ownership (assignee) and inventor information are two principal ways of identifying patents resulting from multinational collaborative inventive activity. The use of patent ownership data to identify such collaborative patent activity is complicated by the wide variance in the form and content of the ownership data, owing to the diverse policies and practices of patent owners and their legal representatives when selecting the owner name to appear on the patent. Although ownership data can be useful as a means of examining the characteristics of identified collaborative patents, it is difficult to use it as the sole means of identifying collaborative patent activity.

Identifying multinational collaborative patents by searching for patents with many inventors residing in different countries also has limitations. These limitations include the further difficulty that such methodology will not identify multinational collaborative activities involving persons not listed on a patent as inventors. Also, collaborative efforts that take place within the framework of a long-term multinational joint venture may not be identified if the inventors all list residences at the location of the joint venture, for example. Furthermore, the use of inventor data focuses attention on the participants in the inventive activity but does not address issues concerning the national or parental ownership of the results. Nevertheless, using inventor residence at the time of patent grant, in conjunction with ownership data, appears to provide a more precise view of multinational collaboration.

For example, a search of 1993 patents for those having both a US resident and a foreign resident inventor revealed 1 499 collaborative patents. By contrast, only 141 patents were identified as having ownership jointly assigned to both a US organisation and a foreign organisation (24 of those patents had a single inventor and 60 also contained both a US and a foreign resident inventor).

Ultimately, inventor and ownership data provide circumstantial evidence about collaborative research activities between countries. Further research, including direct contact with patenting organisations, would be necessary to confirm the extent to which the patent data reflect the true research and development activities of organisations.

CONCLUSION

This study of US patent data finds that US resident inventors are collaborating more frequently with inventors residing abroad, both in terms of total patent output and as a share of total patents granted. This is consistent with other indicators showing an increasing amount of international collaboration in scientific research. These findings support the premise that the inventive process is becoming increasingly multinational in character. While patent data constitute a valuable source of information about the international globalisation of technology, the data are not without their limitations. Because this is a relatively new tool, the appropriate use of patent data continues to be examined. Many of the limitations and strong points of patent data have been well established and have become, themselves, the subjects of discussion and study. This is necessary and beneficial. It is one of the TAF programme goals to advance and facilitate efforts to learn more about the legitimate implications of patent statistics. Many factors need to be considered when analysing these statistics. Among them are the variation in importance and degree of invention of patents. Also, the propensity to patent, as opposed to relying on patent alternatives (such as lead time in the market place and trade secrets) may vary over time, within an industry and among industries. These factors and others may well affect any use of, or conclusions drawn from, patent statistics.

NOTES

1. Special thanks are extended to Wes Gewehr, Administrator for Information Dissemination, and Jane Myers, Director of the Office of Information Products Development (OIPD), who provided managerial support for the completion of this report, and to Jim Declerk of the OIPD Product Development Division, who provided technical support and programming assistance.
2. APS data reflect the content of the printed patent. APS data are also available for sale to the public and commercial vendors.
3. This may also extend to situations where a firm is "trading as" or "also known as" another name. While only the first name appears on the patent record, both names are recorded in the USPTO's patent assignment database. Patents are considered personal property and can be sold to others, mortgaged, bequeathed or otherwise reassigned. This is not usually a problem for analyses involving ownership at the time of issue, as patents are not likely to be reassigned within six months of issue. However, analyses of current patent ownership are more likely to be affected by subsequent ownership changes. When known, changes are recorded in the assignment database, but owners are not required by law to report changes.
4. The practices of some representatives can at times produce undesirable results. In one case, a firm's legal representative mistakenly omitted the assignee's name from the face of several patents.
5. A more accurate percentage of total collaborative patents assigned to these organisations would be slightly lower. A small amount of double-counting resulted from multiple assignees on patent records. Each patent was counted once for each assignee. Relatively few US patent records show multiple assignees. Of the 10 814 patents with international collaboration, only 397 patents (3.6 per cent) had multiple assignees at the time of issue. Thus, this double-counting introduced a rather small amount of error.
6. Corporate directories included: *America's Corporate Families* (Dun and Bradstreet, 1994); *Directory of Corporate Affiliations* (National Register Publishing, 1994); *Directory of Foreign Firms Operating in the United States* (Uniworld Business Publications, Inc., 1992); *Canadian Key Business Directory* (Dun and Bradstreet, 1994); *Standard and Poor's Register of Corporations, Directors and Executives* (McGraw-Hill, 1994); and *Corporate Technology Directory* (Corptech, 1993). Patent activity by US colleges and universities is reported annually by the TAF Report, *US Universities, 1969 to (Present)*, prepared annually for the National Science Foundation.
7. The calculation of the index of collaboration is similar to a chi-square calculation. The index for each organisational category was calculated by multiplying the total number of patents assigned to entities in that category by the total number of collaborative patents assigned to all entities in all categories. This product was then divided by the total number of patents assigned to all entities in all organisations to determine the number of expected collaborative patents. The final index value is calculated by dividing the actual number of collaborative patents by the number of expected collaborative patents.

REFERENCES

ARCHIBUGI, D. and M. PIANTA (1992), "Specialisation and Size of Technological Activities in Industrial Countries: The Analysis of Patent Data", *Research Policy*, Vol. 21, pp. 79-93.

ASPDEN, H. (1983), "Patent Statistics as a Measure of Technological Vitality", *World Patent Information*, Vol. 5, pp. 170-173.

HAGEDOORN, J. (1990), "Organisational Modes of Inter-firm Co-operation and Technology Transfer," *Technovation*, Vol. 10, pp. 17-30.

HOWELLS, J. (1990), "The Internationalisation of R&D and the Development of Global Research Networks", *Regional Studies*, Vol. 24, pp. 495-512.

NATIONAL SCIENCE BOARD (1993), *Science & Engineering Indicators – 1993*, US Government Printing Office (NSB93-1), Washington, DC.

OFFICE OF TECHNOLOGY ASSESSMENT AND FORECAST (OTAF) (1984), "Review and Assessment of the OTAF Concordance Between the US Patent Classification and the Standard Industrial Classification System: Final Report", Washington, DC.

PATEL, P. and K. PAVITT (1991), "Large Firms in the Production of The World's Technology: An Important Case of Non-Globalisation, *Journal of International Business Studies*, Vol. 22, pp. 1-21.

SCHOTT, Thomas (1994), "Collaboration in the Invention of Technology: Globalisation, Regions, and Centres", *Social Science Research*, Vol. 23, pp. 23-56.

ASSESSING DIFFERENT TYPES OF PATENT DATA FOR DESCRIBING HIGH-TECHNOLOGY EXPORT PERFORMANCE

by

Hariolf Grupp, Gunnar Münt and Ulrich Schmoch
Fraunhofer Institute for Systems and Innovation Research (FhG-ISI), Karlsruhe, Germany

INTRODUCTION

This chapter tests different patent databases used to measure innovative activity in high-technology trade among the three leading economic blocs – North America, Europe and Japan. It pays special attention to the problem of identifying the most suitable basis for analysing the internationally relevant technology that has the greatest impact on international trade flows. With the use of a product-oriented definition of high technology, it compares several domestic and international patent samples with the respective trade indicators. Unlike studies that use current state-of-the-art measurement techniques, the method used here moves beyond traditional specialisation indicators to indicators of absolute technological advantages – patent and market shares. It thus develops a new, internationally applicable, patent database: the "triad patent model", which combines the advantages of the US patent database with the now largely available databases from other countries. The model has been successfully used to test the relationship between international technological activity and trade flows among the three economic blocks.

WHAT IS MEANT BY HIGH TECHNOLOGY?

International technological competitiveness is becoming an ever more important issue in international trade negotiations as well as in national economic policy debates (Porter, 1990). Competition on world markets is increasing, especially among the three leading economic blocs, because national wealth depends more and more on export trade flows. New technology is a characteristic feature of this competition, since, to a large degree, it determines success on world markets. As technologies are now created through complex innovation processes, we have to seek the various (institutional and geographical) sources of those technologies. A focus on high technology offers a way to explain world market patterns through differences in technology creation, since the link between technology and trade patterns is rather tight in high-technology product groups (Dosi *et al.*, 1990). Then the focus shifts to identifying and measuring differences in the technological potentials of countries.

Traditionally, national characteristics were thought to shape technology creation; in other words, national strengths and weaknesses depended on the efficiency of mechanisms within national systems of innovation (Lundvall, 1992; Nelson, 1993). In the wake of the increasing globalisation of the world economy, however, national technical change may be outweighed by global technological developments or the increasing share of multinational corporations in international production (Fagerberg, 1994). Therefore, we will explore differences in countries' technological potential and then ask whether national characteristics still dominate technological change. First, however, it is necessary to define high technology. This is no easy matter, since "high technology" is, to a significant extent, a public and political perception and therefore not readily amenable to precise definition in terms of economics or innovation research (Johnston and Hartley, 1986).

According to some authors, high technology is defined by R&D intensity; in this case, R&D expenditure divided by turnover is the basis for classification. However, this definition is problematic, as it divides industries, and therefore all the products they manufacture or services they provide, into two categories: high technology or not. This is the definition used by the OECD, the US Department of Commerce, and other public bodies. Yet, the present techno-economic system is characterised by the increasing importance of technology fusion or spillover effects (Kodama, 1991; Grupp, 1991, 1992 and 1994*a*; see also other references cited by these authors). However, industries do not represent homogeneous technologies. A way out of these difficulties is to use product-based R&D intensities (Abbott *et al.*, 1989; Legler, 1987; Grupp, 1991), a procedure that is still heavily debated (United States International Trade Commission, 1990).

Following the revision of the International Trade Classification, the more traditional definition of high technology may no longer be appropriate, at least at the international level, since 1988. The Fraunhofer Institute for Systems and Innovation Research (FhG-ISI) has therefore designed a new list of R&D-intensive products (Legler *et al.*, 1992), which is modelled on an earlier effort by the Lower Saxony Institute for Economic Research (NIW), which was, in turn, based on the previous International Trade Classification (Legler, 1987).

Table 1 lists the product groups considered as high technology; they are defined in terms of the Standard International Trade Classification (SITC), Rev. 3. The R&D expenditure and turnover of individual firms in major OECD countries at the end of the 1980s were sorted, along with their R&D intensity. Owing to the lack of internationally comparable data, national data were used. Company R&D statistics are available for the United States (Standard & Poor's Compustat Services) and for Germany (Schwitalla, 1993), and product-based statistics are available for Japan,

◆ Table 1. ***List of R&D-intensive products and government intervention***

No.	SITC Rev. 3	Product group (unofficial terms)	PP	GP	GS	BE	PR	ER
1	516	Advanced organic chemicals		░				░
2	525	Radioactive materials	░	░	░	░		░
3	541	Pharmaceutical products		░		░	░	░
4	575	Advanced plastics						░
5	591	Agricultural chemicals				░	░	░
6	714	Turbines and reaction engines		░	░			
7	718	Nuclear, water, wind power generators		░	░	░	░	░
8	752	Automatic data processing machines		░	░			
9	764	Telecommunications equipment		░		░	░	
10	774	Medical electronics		░				
11	776	Semiconductor devices		░		░		
12	778	Advanced electrical machinery		░				
13	792	Aircraft and spacecraft		░		░		
14	871	Advanced optical instruments		░		░		
15	874	Adv. measuring instruments		░		░		
16	891	Arms and ammunition		░		░		
17	266	Synthetic fibres						
18	277	Advanced industrial abrasives						
19	515	Heterocyclic chemistry						■
20	522	Rare inorganic chemicals						
21	524	Other precious chemicals						■
22	531	Synthetic colouring matter						■
23	533	Pigments, paints, varnishes						
24	542	Medicaments		■		■		
25	551	Essential oils, perfumes, flavours						■
26	574	Polyethers and resins						■
27	598	Advanced chemical products						■
28	663	Mineral manufactures, fine ceramics						
29	689	Precious non-ferrous base metals	■					
30	724	Textile and leather machinery						
31	725	Paper and pulp machinery						
32	726	Printing and bookbinding machinery		■				
33	727	Industrial food-processing machines						■
34	728	Advanced machine tools						
35	731	Machine tools working by removing...						
36	733	Machine tools without removing...						
37	735	Parts for machine tools						
38	737	Advanced metalworking equipment						
39	741	Industrial heating and cooling goods						■
40	744	Mechanical handling equipment						
41	745	Other non-electrical machinery						
42	746	Ball and roller bearings						
43	751	Office machines, word processing		■				
44	759	Advanced parts for computers		■	■			
45	761	Television and video equipment				■		
46	762	Radio broadcast, radiotelephony goods				■		
47	763	Sound and video recorders				■		
48	772	Traditional electronics						
49	773	Optical fibre and other cables		■				
50	781	Motor vehicles for persons				■		■
51	782	Motor vehicles for goods transport						■
52	791	Railway vehicles		■		■	■	
53	872	Medical instruments and appliances		■				
54	873	Traditional measuring equipment						
55	881	Photographic apparatus and equipment						
56	882	Photographic and cinematographic supplies						
57	884	Optical fibres, contact, other lenses						

PP : Large share of public production
GP: Considerable government procurement
GS: Noticeable government subsidies
BE : Barriers to market entry or production quantity
PR : Strong price regulation
RE : Environmental regulation

░ Leading edge technology

■ High-level technology

Source: Grupp (1994*c*).

the United States, Germany, Italy, Sweden and the Netherlands (see Grupp, 1991). Frequency analysis was used to set the cut-off for high technology: all products with R&D intensities above the industry average of about 3.5 per cent of turnover are considered to be high-technology products.

The list offers the advantage of dividing the R&D-intensive sector into two parts and differentiating between leading-edge products and products with high-level technology. Frequency analysis of R&D intensity was again used to set the threshold, which falls at 8.5 per cent of turnover.

A similar product-based high-technology list was presented by Amendola and Perrucci (1990, 1992). Stressing the limitations of sectoral classifications of R&D intensity, these authors based their selection of products on the NIMEXE nomenclature used by the European Union (EU), which is more disaggregated than the SITC III classification used here. They conducted specific inquiries in manufacturing firms and asked experts to evaluate products on the basis of a number of parameters including R&D intensity, the degree of automation in production, characteristics of product use and product life cycle. In this way, they defined 254 high-technology products at a six-digit level of disaggregation. A comparison of these products to the definition used for Table 1 shows that more than 70 per cent fall into the category of leading-edge technology and less than 30 per cent into high-level products. Thus, this newly designed definition of high technology may be more or less equated with highly R&D-intensive production.

Why such a distinction between very R&D-intensive (leading-edge technology) and (above-average) high-technology (high-level) products? The definition may be turned around: a very high percentage of expenditure on R&D signifies low turnover expectations. Indeed, for every million spent on R&D in leading-edge technology, the average turnover is less than 12 million, while a typical figure for high-level technology is 30 million and more for each million invested in R&D (Grupp, 1991). But again, why is the turnover expectation for leading-edge technology lower than for the other high-technology category?

Table 1 also classifies these products, somewhat impressionistically, in terms of six modes of government intervention. Some product groups show a large share of public production (PP), while others show strong government procurement (GP). Energy generation, aircraft and spacecraft. among others, receive government subsidies (GS) in many countries. Other product groups face barriers to market entry (BE) or to production or at least export quantities; this is the case for the pharmaceutical sector, for example, which requires approval by food and drug administrations or health care authorities. Consumer electronics, along with cars, steel and textiles, have seen at various times the application of voluntary export restraints, technical barriers to trade (*e.g.* colour television systems), and the like (Cawson and Holmes, 1991, p. 1973). Strong price regulation (PR) exists at least for railway and telecommunication services, and environmental regulation (ER), a more recent form of government intervention, is becoming more and more pervasive.

Thus, as Table 1 indicates, government intervention is greater for leading-edge technology than for high-level technology. Leading-edge technology includes many products that are subject to protectionism, such as aeronautics and aerospace or nuclear energy, whereas most high-level products are mass consumption products traded internationally more or less under free market conditions. Thus, the distinction between the two high-technology categories seems to be meaningful, at least in terms of a model for further study of the various aspects of national science and technology production and its impact on competitive trade advantage. Analogous discussions of high technology, procurement, trade barriers and regulation in the Single European Market after 1992 may also be found in Buigues *et al.* (1990).

THEORETICAL BACKGROUND AND INDICATORS

Competitiveness is an elusive concept. For some years now, it has been commonly understood that competitiveness is largely influenced by a nation's ability to generate and sell high-technology products at home and abroad. This school of economic thought essentially derives from the work of Schumpeter (1912, p. 99). Foreign trade dynamics based on Schumpeterian arguments were developed in particular by Posner (1961) and Lorenz (1967), among others. According to this thinking, a country rich in innovations will create a technology gap with competing countries that are less active or less successful in producing innovations. This gap will determine their technological competitiveness and thus also what goods they trade.

The focus of attention is generally the relationship between the generation of high-technology products, international trade, and economic growth. While various economic indicators illustrate economic growth and subsequent welfare gains, innovation and trade indicators cannot yet account for the impact of high-technology production on international trade. The measurement of international technological activity depends almost entirely on patent statistics. Thanks to the contributions of Pavitt and Soete (1980), Soete (1981, 1987), Legler (1987), Dosi *et al.* (1990), Legler *et al.* (1992), and Grupp *et al.* (1992), there is, however, a great deal of evidence for a strong relationship between an economy's technological and trade specialisation.

However, technological competitiveness is quite different from traditional neoclassical comparative advantage. Following Dosi and Soete (1988), we regard comparative national advantages as being dominated by absolute advantages, which act as the main determinant of international trade when countries have different technological potentials. Consequently, we have to look at countries' market shares in different product groups and verify the relationship between countries' technological activity in certain high-technology fields and their corresponding world market share. On the basis of the distinction between absolute (technological or structural) and comparative (price-related) advantages, we use output indicators to measure technological activity and world market success. As far as success on world markets is concerned, several indicators are available. World export shares, which are generally consistent with the theoretical concept of absolute advantages, appear to be the most appropriate indicator. It is defined as follows:

$$WES_{ikt} = (ex_{ikt}/\Sigma_k \, ex_{ikt})$$

where ex_{ikt} denotes the exports of country i in product group k at time t. The WES indicator measures the level of economic activity a country has reached on the world market for a special product. A larger share of world exports suggests that the country has a higher level of absolute advantage than other countries. On the basis of the argument presented above, we suppose that higher levels of absolute advantage on certain world market segments (product groups) fall together with higher levels of technological activity as evidenced by output of patents. Thus, we take the absolute number of a country's "triad patents" in a given technological field and relate it to the number of all triad patents in this field. The world patent share is denoted by WPS_{ikt}.

Further, we want to know how technological productivity in high-technology areas impinges on competitiveness and relates to the world market shares in the (more regulated) leading-edge and the (less regulated) high-level areas. Therefore, we adjust the patent indicator for the size of a country's labour force. This indicator is defined as follows:

$$TPL_{ikt} = (TP_{ikt}/L_{it})$$

TPL_{ikt} denotes the triad patents per labour force (L_{it}) in country i and technological field k at time t. Triad patents are preferred to domestic patent counts so as to adjust for different regional propensities to patent (to be described in detail below). Adjusting for the size of the labour force reveals patent intensities (or "technological productivity" per worker). We are interested in whether single countries can increase their world market shares in high-technology areas by specialising in specific high-technology fields. Thus, if the model of absolute technological competitiveness applies to all high-technology product categories (or technological areas), we should also find a positive correlation between world market shares and technological intensity. However, before presenting our results, we discuss some related problems.

NATIONAL CONTROL OF R&D OR GLOBALISATION?

Technoglobalism puts pressure on national technology and industrial policies. More than in the past, national policies face competition from both supra-national bodies (such as the European Commission or NAFTA bodies) and regional activities (such as the state and province policies in the United States, Germany, Italy or Spain) and local activities (such as campus-based incubators). Further, direct investment by multinational firms, global sourcing, networking and the "mobile" location of R&D establishments of firms have advanced more rapidly than the related government policies.

This "globalisation" would suggest including all countries of the world in such an analysis. However, it is more correct, theoretically, to consider only bilateral trade relations. After all, it is by no means certain – indeed, there is some evidence to the contrary – that the advanced economies always supply products of the same high technological content to each other. This makes consideration of trade relations within groups of similar economies preferable (Grupp 1991, p. 281). As a compromise solution, we analyse the 18 larger OECD economies (see Figures 1 and 2), unlike Daniels (1993), who covered 52 countries.

The task is to use patents to measure the disembodied knowledge of countries. When bringing patent and trade data together, there is one main caveat relating to the extent to which input and output variables reflect R&D processes under national control (Grupp *et al.*, 1992). For R&D expenditure, *i.e.* input into R&D, this does not

◆ Figure 1. ***Relation of international "triad" patents (per labour force 1986-88) and world export share (1990) for high-level technology***

WES (i, k, 1990) = b + m TPL (i, k, 1986-88)

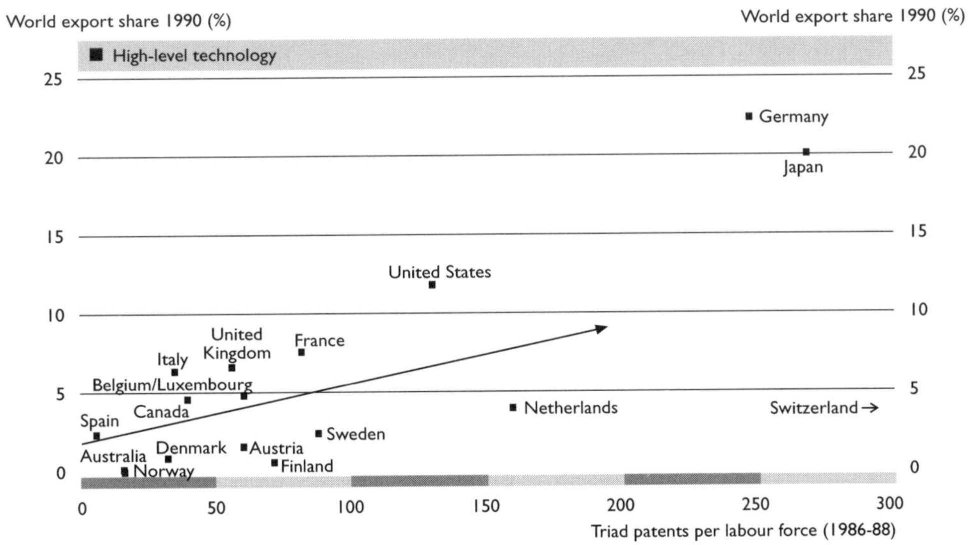

Source: FhG-ISI.

◆ Figure 2. **Relation of international "triad" patents (per labour force 1986-88) and world export share (1990) for leading edge technology**
WES (i, k, 1990) = b + m TPL (i, k, 1986-88)

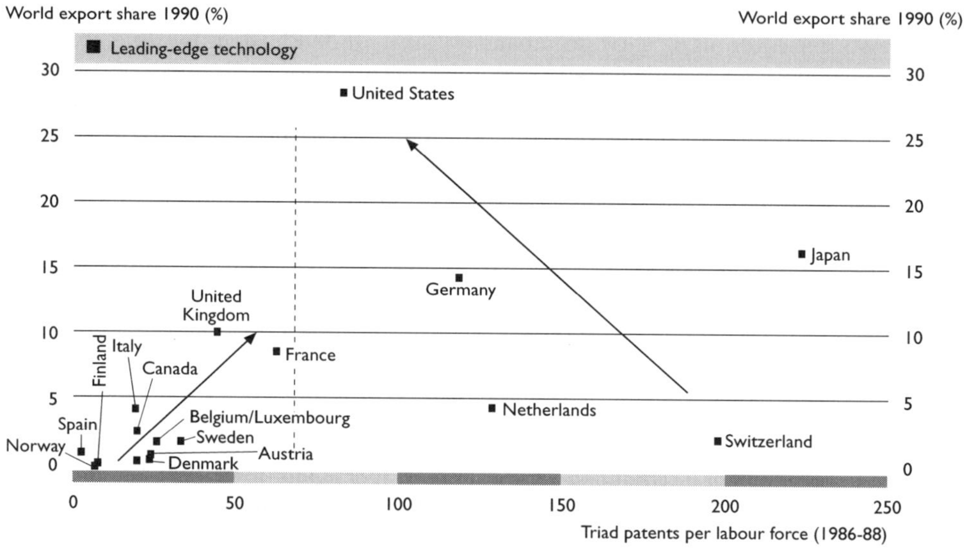

Source: FhG-ISI.

appear to present a problem, as the flow of funds to and from foreign sources is negligible in the OECD countries. More problematic is the strategic control of the technology generated by patents (Grupp *et al.*, 1992). In most cases, inventors do not decide when, to what extent, or for which markets or products their inventions are used. Nor, insofar as they are employees of commercial firms, do they generally have a say in licencing, selling or abandoning patents. Belgian technology, for instance, is controlled to a large degree by non-Belgian companies. Patel and Pavitt (1992) reported that nearly 40 per cent of Belgian patenting in the United States is attributable to large foreign-controlled firms, whereas for Western Europe the comparable figure is 6 per cent, on average. Belgium had the largest share of national patents generated by foreign capital in a sample of 11 countries.

Even more problematic is the foreign trade indicator WES. It is by no means certain that high-technology exports are built on technology and science indigenous to the exporting country. As a public good, "R&D results" are highly mobile. In particular, large multinational corporations may produce and export from other countries than those where the scientific and technological achievements originate.[1] The level of high-technology trade advantage will thus depend not only on the existence of scientific or technology gaps, but also on the number and nature of high-technology firms that manufacture in that country, and thus on the level of opportunity, cumulativeness and appropriability (*i.e.* the inter-firm technological capabilities) of the local export sector (van Hulst *et al.*, 1991).

A study of major OECD countries includes EU countries, which creates another problem. There are increasing amounts of intra-EU trade (see Grupp and Soete, 1993, pp. 58-64). At least in the information industry, trade diversion seems to have taken place along with the recent phases of economic integration within the Single Market. If we consider this as regional trade, the international share of some countries will be less than 50 per cent of their exports. For other OECD countries, such as the United States or Japan, the regional and overseas shares in exports are quite different.

These reservations should be kept in mind when assigning both patent data and foreign trade data to individual countries.

A SIMPLE TRIAD MODEL FOR OECD COUNTRIES

Private or corporate research generally produces patents rather than academic publications (Dasgupta and David, 1987; Grupp, 1990; OECD, 1994). Patent statistics are an accepted output indicator of codified knowledge from strategic and applied research and industrial development. As patent applications are legal documents valid in a single country, many domestic priority patent applications are "duplicated" abroad. Selection of patent data from a single patent office, therefore, does not always yield an indicator that is representative of the world output of inventions. Using the patent classification of the United States, it is possible to match patents to trade categories (Legler *et al.*, 1992); thus, we examined the classifications disclosed in patents issued by the United States Patent and Trademark Office further to break down technology.

Among others, Legler *et al.* (1992) and Amendola *et al.* (1993) found, in different types of investigations, that because of the cumulative nature of innovation, patent statistics precede international trade by about three years. Here, patents granted for the invention years 1986 through 1988 were selected, and may be regarded as proxies for corporate attempts to protect goods produced in 1990. Accordingly, trade data from 1990 are used.

The patents granted in the United States, whose product classification was used, may serve as a benchmark for an analysis of the bilateral trade between any one OECD country and the United States. It is questionable, however, whether international trade (notably within the EU and between the EU and Japan) is reflected in intellectual property rights in the United States. From a theoretical standpoint, therefore, this is an insufficient basis. Therefore, we used another definition of relevant intellectual property rights that is significant for international trade.

As patent duplications can be traced and matched to each other, it is possible to define "patent families" that are grouped around one invention and have property rights in various countries. In the following analysis, only the inventions for which a patent was filed at least in the United States, Japan and the European Patent Office are taken into consideration. On this basis, we apply the "triad" model. As in the United States a patent was not only filed but also granted, the classification may be taken from the US equivalent and matched to the product group of potential use. Thus, these patents were assigned to either high-level or leading-edge products by a patent-to-sales concordance (Legler *et al.*, 1992), instead of using US patent data alone (the standard procedure: see Frame, 1991; and Dosi *et al.*, 1990). In this way, we can consider that we cover world technology output. However, it is not possible to describe this method in detail here (see Grupp, 1994*b*).

In order to test the usefulness of the triad concept, we examined the world market share for all manufactured products in relation to various sets of patent data, among them the triad patents. Further, we assigned the US patents to inventor countries (residence of inventors) as well as to assignee countries (residence of owner company) to gain an impression of the effects of globalisation. The first set (USINV) represents the country of origin, the second the firm/country controlling the technology (USASS). Next, we selected Japanese patent applications [at the Japanese office Tokyo cho (TC)], patent applications at the German Patent Office (DPA), and patent applications at the European Patent Office (EPO).

From these data sets, we constructed three more patent samples (all sets being adjusted to 1986-88 priority years, including US patents granted; we did not use patent publication dates[2]). We merged all domestic patent applications for the OECD countries (DOM), we added the EPO patents with destination Germany to the DPA data (and denoted this set as GE for Germany/Europe), because in Germany legal protection is assured both ways, and then finally selected the triad patents (TP) as described above.

EMPIRICAL RESULTS

By linear regression of the data sets (in each case, the world patent share per set) with the world export share WES in manufactured goods only we obtained the results shown in Table 2. Two sets of patents are highly correlated with absolute export performance: the triad patents (as expected) and the European data. Since the EPO requires fees considerably higher than those charged by the European national patent offices, the application is only cost-effective if the application is

Table 2. **Regression results for world export share (1990) and seven patent samples (priority dates 1986-88)**

Patent sample	R^2	t	Degrees of freedom
USINV	0.44	3.46**	15
USASS	0.43	3.36**	15
TC	0.15	1.62	15
DPA	0.45	2.57*	8
EPO	0.88	10.41***	15
DOM	0.27	2.38*	15
GE	0.64	3.79**	8
TP	0.62	4.90***	15

*** Highly significant at the 0.1% level.
** Significant at the 1% level.
* Weakly significant at the 5% level.
Source: Author.

designated for three or more countries under the European Patent Convention. This seems to indicate a process for selecting the most important patents for world markets similar to that achieved by the triad model.

The US patent data explain trade advantages reasonably well (significant correlation) and there is almost no difference between the inventors' countries and the assignees' countries; this works against any strong distortion effect from globalisation. The correlation is reasonably good for all patents with protection in Germany (GE), but for patent documents at the German Patent Office alone (DPA), *i.e.* neglecting access to protection through the EPO, the correlation is weak. Applications at domestic patents do not explain world trade well, nor do patents in Japan.

From this we conclude that – in line with the evolutionary concept of absolute advantages – higher levels on certain world market segments, *i.e.* manufactured goods, parallel higher levels of technological activity as evidenced by patent output. However, because protection is limited to national markets, an analysis of the trade advantage of high technology should be based either on European or triad patents. As no concordance of the International Patent Classification, which is applied by the EPO, to the SITC is available, we chose to match triad patents and trade categories using the US product classification.

A linear regression was used to investigate how technological productivity relates to world market shares in high technology:

$$WES_{ik1990} = b + m \cdot TPL_{ik(1986\text{-}88)}$$

For k we selected:

– all high-technology products and related patents as defined in Table 1;

– all leading-edge products and related patents as defined in Table 1;

– all high-level products and related patents as defined in Table 1.

Figures 1 and 2 provide an overview of the technological productivity of OECD countries.[3] The correlation between international "triad" patents with potential application in high-level products and the respective trade data is weak, but it exists (a linear regression with $R^2 = 0.29$ is significant at the 5 per cent level, Figure 1). The most notable disparity is observed for Switzerland, a very productive country (in terms of technology) but one that is isolated in the geographical centre of the European economy, as it does not want to join the European Union but is surrounded by member countries.[4] Yet, there is no significant correlation for the remaining patents with codified knowledge for potential use for leading-edge technology (a linear regression with $R^2 = 0.23$ is not significant at the 5 per cent level, Figure 2). If we bring both high-technology sectors together, the correlation holds on a weak but significant level ($R^2 = 0.30$, no figure).

Technology production, as expressed through intellectual property rights, seems to have a very important function for structural advantages in the more market-oriented parts of high technology. Despite the above-mentioned reservations about the control of technology through patents, which can easily cross borders, national factors are strong enough overall to maintain a strong coupling between the local source of the invention and the national share in world trade.

For leading-edge technology the situation is more varied. As a guide, we inserted two lines in Figure 2, which are not regression lines. With too few patents, a nation has no chance to compete on international technology markets. With more technology output, as in France and the United Kingdom, absolute advantages on international markets accrue. But above a certain threshold, the situation is distorted by factors such as protectionism and regulation, and government-protected leading-edge markets are controlled by means other than intellectual property rights. As a result, the more direct relation between technology produced and success in trade is vanishing. The United States is the world leader in leading-edge exports at the medium technology production level. Germany, the Netherlands, Switzerland and Japan try hard to catch up in leading-edge technology – but with little success in trade so far.

CONCLUSIONS AND AN OUTLOOK ON NATIONAL SPECIALISATION

This investigation reveals different patterns of technology production and competitive trade. National systems of innovation continue to be very strong, and technological developments cannot be viewed as a global phenomenon, independent of a national environment. The globalisation of modern technology, as important as it may be, does not relieve national technology policy of responsibility for local R&D activities and innovative industrial sites.

This study contains an initial analysis of international trade based on disaggregation procedures different from those used for industries. It seems useful to divide the *high-technology sector* into two parts in order to make an in-depth analysis of national strengths and weaknesses. The rationale behind this division into two R&D-intensive product groups is that there are, on the one hand, the *high-level products* for which there are expectations of a relatively good return on R&D investment; here, technological production originating from R&D activities in local industry plays a role and is decisive for international competitiveness. Scientific achievements are of medium importance here. On the other hand, there are the *leading-edge technologies* with expectations of only moderate return on R&D investment; here, factors other than technology guarantee international success, and there is stronger government intervention in terms both of R&D and of procurement and regulation. Here, national technology production does not fully result in competitiveness.

It is a severe limitation of this study that it tests the relationship of innovative proxies and countries' exports by cross-sectional data (allowing for a fixed time lag). Although the correlation between the two classes of high technology in the OECD nations and their respective export performances are robust, this type of test does not give a proper account of how technical change actually determines international competitiveness (see Amendola *et al.*, 1993). Bilateral comparison and analysis of the development of flows of trade within or between trade blocs show the extent to which the large industrialised nations have specialised in different product areas and how these priorities have developed over time. There clearly is a world-wide division of labour in the high-technology area (Legler *et al.*, 1992).

The analytical instruments introduced in this paper seem to be well suited for broader application in economic analysis and for technology policy. Policies aiming at improvements in competitiveness ought to focus primarily upon the factors that influence international trade. National technology production seems to have been a major element of market success up to now, at least for high-level technologies, despite increasing intra-firm flows of knowledge.

However, the lack of a concordance between the international trade (SITC III) and the international patent classification (IPC) is a serious one. Here, we used the US Patent and Trademark Office concordance, which does not fit the international classifications. The US patent system provides unique data sources in this respect and is more advanced than other countries or the OECD as a whole.

Applying the triad model and thus original statistical data from countries other than the United States requires complicated classification transactions which may lead to serious errors. Efforts should be devoted to working out a patent-to-trade concordance directly at international level. Case studies in information technology have been performed (Grupp and Soete, 1993) that show convincingly that such an attempt can be successful, provided that scientists working on this subject dispose of the necessary economic and technological knowledge.

NOTES

1. Israel, although not an OECD Member country and not included in this paper, would be the most striking case in this context; see Grupp *et al.* (1992).
2. "Patent dates" are identical to publication dates.
3. Note that Belgium and Luxembourg, which form an economic and monetary union, are treated as a single country. Note also that all German data are for West Germany only.
4. The exceptional disparity between the Swiss position in technology and trade might be in part due to the fact that some important Swiss companies have become large multinationals. Thus the location of technology creation and production may be different.

REFERENCES

ABBOTT, T., R. Mcguchin, P. HERRICK, and L. NORFOLK (1989), "Measuring the Trade Balance in Advanced Technology Products", Centre for Economic Studies, Discussion Paper CES 89-1. Bureau of the Census, US Department of Commerce, Washington, DC.

AMENDOLA, G. and A. PERRUCCI (1990), *Specialisation and Competitiveness of Italian Industry in High-Technology Products*, ENEA, Rome.

AMENDOLA, G. and A. PERRUCCI (1992), "European Patterns of Specialisation in High-Technology Products: A New Approach", *Proceedings of the Joint EC-Leiden Conference on Science and Technology Indicators*, DSWO Press, Leiden.

AMENDOLA, G., G. DOSI, and E. PAPAGNI (1993), "The Dynamics of International Competitiveness", *Weltwirtschaftliches Archiv*, Vol. 129, No. 3, pp. 451-471.

ARCHIBUGI, D. and M. PIANTA (1991), "Specialisation and Size of Technological Activities in Industrial Countries: The Analysis of Patent Data", Consiglio Nazionale delle Ricerche, Istituto di Studi sulla ricerca e documentazione scientifica, Rome.

BUIGUES, P., F. ILZKOVITZ, and J.-F. LEBRUN (1990), "Industrieller Strukturwandel im europäischen Binnenmarkt; Anpassungsbedarf in den Mitgliedstaate", *Europäische Wirtschaft – Soziales Europa*, Commission of the European Communities, Brussels, Luxemburg.

CAWSON, A. and P. HOLMES (1991), "The New Consumer Electronics", in C. Freeman, M. Sharp and W. Walker (eds.), pp. 169-182, *Technology and the Future of Europe: Global Competition and the Environment in the 1990s*, Pinter Publishers, London, New York.

DANIELS, P. (1993), "Research and Development, Human Capital and Trade Performance in Technology-intensive Manufacturers: A Cross-country Analysis", *Research Policy*, Vol. 22, pp. 207-241.

DASGUPTA, P. and P.A. DAVID (1987), "Information Disclosure and the Economics of Science and Technology", in: G. Feiwel (ed.), *Arrow and the Ascent of Modern Economic Theory*, New York University Press, New York.

DOSI, G. and L. SOETE (1988) "Technical Change and International Trade", in G. Dosi, C. Freeman, R. Nelson, G. Silverberg and L.L.G. Soete (eds.), *Technical Change and Economic Theory*, Pinter Publishers, London, New York.

DOSI, G., K. PAVITT, and L.L.G. SOETE (1990), *The Economics of Technical Change and International Trade*, Harvester Wheatsheaf, New York, London.

FAGERBERG, J. (1994), "Is there a Small-Country Disadvantage in High-Tech?", paper presented at the 1994 Münster Conference of the International Joseph A. Schumpeter Society, 17-20 August.

FRAME, J. (1991), "Modelling National Technological Capacity", *Scientometrics*, Vol. 22 (No. 3), pp. 327-339.

GLISMANN, H.H. and E.-J. HORN (1988), "Comparative Invention Performance of Major Industrial Countries: Patterns and Explanations", *Management Science*, Vol. 34, pp. 1169-1187.

GRUPP, H. (1990), "On the Supplementary Functions of Science and Technology Indicators", *Scientometrics*, Vol. 19, pp. 447-472.

GRUPP, H. (1991), "Innovation Dynamics in OECD Countries: Towards a Correlated Network of R&D-Intensity, Trade, Patent and Technometric Indicators", in OECD, *Technology and Productivity: The Challenge for Economic Policy*, Paris, pp. 275-295.

GRUPP, H. (ed.) (1992), *Dynamics of Science-Based Innovation*, Springer Publishers, Berlin, Heidelberg, New York.

GRUPP, H. (1994*a*), "Spillover Effects and the Science Base of Innovations Reconsidered: An Empirical Macroeconomic Approach", Paper presented at the 1994 Münster Conference of the International Joseph A. Schumpeter Society, 17-20 August, also to be published in *Journal of Evolutionary Economics*, 1996.

GRUPP, H. (1994*b*), "The Measurement of Technical Performance of Innovations by Technometrics and its Impact on Established Technology Indicators", *Research Policy*, Vol. 23, No. 2, pp. 175-193.

GRUPP, H. (1994*c*), "Science, High Technology and the Competitiveness of EC Nations", *Cambridge Journal of Economics,* Vol. 19, No. 1, pp. 209-223.

GRUPP, H., S. MAITAL, A. FRENKEL, and K. KOSCHATZKY, (1992), "A Data Envelopment Model to Compare Technological Excellence and Export Sales in Israel and European Community Countries", *Research Evaluation*, Vol. 2, No. 2, pp. 87-101.

GRUPP, H. and L.L.G. SOETE (1993) *Analysis of the Dynamic Relationship between Technical and Economic Performances in Information and Telecommunication Sectors*, 3 vols., Final report Nr 8862 for the European Commission, DG XIII, Karlsruhe and Maastricht

JOHNSTON, R. and J. HARTLEY, 1986. "Formulation and Development of High Technology Indicators", Centre for Technology and Social Change, University of Wollongong.

KODAMA, F. (1991), *Analyzing Japanese High Technologies*, Pinter Publishers, London, New York.

LEGLER, H. (1987), "West German Competitiveness of Technology-intensive Products", in Grupp, H., (ed.), *Problems of Measuring Technological Change*, Verlag TUEV Rhineland, Cologne.

LEGLER, H., H. GRUPP, B. GEHRKE, and U. SCHASSE (1992), *Innovationspotential und Hochtechnologie*, Physica-Springer, Heidelberg, New York.

LORENZ, D. (1967), *Dynamische Theorie der internationalen Arbeitsteilung*, Duncker & Humblot, Berlin.

LUNDVALL, B. (ed.) (1992), *National Systems of Innovation. Towards a Theory of Innovation and Interactive Learning*, Pinter Publishers, London.

NELSON, R.R. (1993), *National Innovation Systems. A Comparative Analysis*, Oxford University Press, Oxford.

OECD (1991), *Basic Science and Technology Statistics*, Paris, and similar OECD sources.

OECD (1994), "Using Patent Data as Science and Technology Indicators: Patent Manual 1994", OCDE/GD(94)114, Paris.

PATEL, P. and K. PAVITT (1992), "Large Firms in the Production of the World's Technology: An Important Case of "Non-globalization", *The Journal of International Business Studies*, Vol. 22, No. 1, pp. 1-21.

PAVITT, K. and L.L.G. SOETE (1980), "Innovative Activities and Export Shares: Some Comparisons between Industries and Countries", in K. Pavitt (ed.), *Technical Innovation and British Economic Performance*, pp. 38-66, Macmillan, London.

PORTER, M.E. (1990), *The Competitive Advantage of Nations*, Macmillan, London.

POSNER, M. V. (1961), "Technical Change and International Trade", *Oxford Economic Papers*, Vol. 13, pp. 323-341.

SCHUMPETER, J. A. (1912), *Theorie der wirtschaftlichen Entwicklung,* Duncker & Humblot, Berlin, 6th edition, 1964.

SCHWITALLA, B. (1993), *Messung und Erklärung industrieller Innovationsaktivitäten mit einer empirischen Analyse für die westdeutsche Industrie,* Physica-Verlag Springer, Heidelberg.

SOETE, L.L.G. (1981), "A General Test of Technological Gap Trade Theory", *Weltwirtschaftliches Archiv*, Vol. 117, pp. 638-660.

SOETE, L.L.G. (1987), "The Impact of Technological Innovation on International Trade Patterns: The Evidence Reconsidered, *Research Policy*, Vol. 16, pp. 101-130.

UNITED STATES INTERNATIONAL TRADE COMMISSION (1990), "Identification of US Advanced-Technology Manufacturing Industries for Monitoring and Possible Comprehensive Study", Report to the Committee on Finance, United States Senate, on Investigation No. 332-294. US/TC Publication 2319, Washington.

VAN HULST, N., N.R. MULDER and L.L.G. SOETE (1991), "Export and Technology in Manufacturing Industry", *Weltwirtschaftliches Archiv*, Vol. 127, pp. 246-264.

MAIN SALES OUTLETS OF OECD PUBLICATIONS
PRINCIPAUX POINTS DE VENTE DES PUBLICATIONS DE L'OCDE

ARGENTINA – ARGENTINE
Carlos Hirsch S.R.L.
Galería Güemes, Florida 165, 4° Piso
1333 Buenos Aires Tel. (1) 331.1787 y 331.2391
Telefax: (1) 331.1787

AUSTRALIA – AUSTRALIE
D.A. Information Services
648 Whitehorse Road, P.O.B 163
Mitcham, Victoria 3132 Tel. (03) 9873.4411
Telefax: (03) 9873.5679

AUSTRIA – AUTRICHE
Gerold & Co.
Graben 31
Wien I Tel. (0222) 533.50.14
Telefax: (0222) 512.47.31.29

BELGIUM – BELGIQUE
Jean De Lannoy
Avenue du Roi 202 Koningslaan
B-1060 Bruxelles Tel. (02) 538.51.69/538.08.41
Telefax: (02) 538.08.41

CANADA
Renouf Publishing Company Ltd.
1294 Algoma Road
Ottawa, ON K1B 3W8 Tel. (613) 741.4333
Telefax: (613) 741.5439

Stores:
61 Sparks Street
Ottawa, ON K1P 5R1 Tel. (613) 238.8985
211 Yonge Street
Toronto, ON M5B 1M4 Tel. (416) 363.3171
Telefax: (416)363.59.63

Les Éditions La Liberté Inc.
3020 Chemin Sainte-Foy
Sainte-Foy, PQ G1X 3V6 Tel. (418) 658.3763
Telefax: (418) 658.3763

Federal Publications Inc.
165 University Avenue, Suite 701
Toronto, ON M5H 3B8 Tel. (416) 860.1611
Telefax: (416) 860.1608

Les Publications Fédérales
1185 Université
Montréal, QC H3B 3A7 Tel. (514) 954.1633
Telefax: (514) 954.1635

CHINA – CHINE
China National Publications Import
Export Corporation (CNPIEC)
16 Gongti E. Road, Chaoyang District
P.O. Box 88 or 50
Beijing 100704 PR Tel. (01) 506.6688
Telefax: (01) 506.3101

CHINESE TAIPEI – TAIPEI CHINOIS
Good Faith Worldwide Int'l. Co. Ltd.
9th Floor, No. 118, Sec. 2
Chung Hsiao E. Road
Taipei Tel. (02) 391.7396/391.7397
Telefax: (02) 394.9176

CZECH REPUBLIC – RÉPUBLIQUE TCHÈQUE
Artia Pegas Press Ltd.
Narodni Trida 25
POB 825
111 21 Praha 1 Tel. (2) 2 46 04
Telefax: (2) 2 78 72

DENMARK – DANEMARK
Munksgaard Book and Subscription Service
35, Nørre Søgade, P.O. Box 2148
DK-1016 København K Tel. (33) 12.85.70
Telefax: (33) 12.93.87

EGYPT – ÉGYPTE
Middle East Observer
41 Sherif Street
Cairo Tel. 392.6919
Telefax: 360-6804

FINLAND – FINLANDE
Akateeminen Kirjakauppa
Keskuskatu 1, P.O. Box 128
00100 Helsinki

Subscription Services/Agence d'abonnements :
P.O. Box 23
00371 Helsinki Tel. (358 0) 121 4416
Telefax: (358 0) 121.4450

FRANCE
OECD/OCDE
Mail Orders/Commandes par correspondance:
2, rue André-Pascal
75775 Paris Cedex 16 Tel. (33-1) 45.24.82.00
Telefax: (33-1) 49.10.42.76
Telex: 640048 OCDE

Internet: Compte.PUBSINQ @ oecd.org

Orders via Minitel, France only/
Commandes par Minitel, France exclusivement :
36 15 OCDE

OECD Bookshop/Librairie de l'OCDE :
33, rue Octave-Feuillet
75016 Paris Tel. (33-1) 45.24.81.81
(33-1) 45.24.81.67

Dawson
B.P. 40
91121 Palaiseau Cedex Tel. 69.10.47.00
Telefax : 64.54.83.26

Documentation Française
29, quai Voltaire
75007 Paris Tel. 40.15.70.00

Economica
49 rue Héricart
75015 Paris Tel. 45.78.12.92
Telefax : 40.58.15.70

Gibert Jeune (Droit-Économie)
6, place Saint-Michel
75006 Paris Tel. 43.25.91.19

Librairie du Commerce International
10, avenue d'Iéna
75016 Paris Tel. 40.73.34.60

Librairie Dunod
Université Paris-Dauphine
Place du Maréchal de Lattre de Tassigny
75016 Paris Tel. 44.05.40.13

Librairie Lavoisier
11, rue Lavoisier
75008 Paris Tel. 42.65.39.95

Librairie des Sciences Politiques
30, rue Saint-Guillaume
75007 Paris Tel. 45.48.36.02

P.U.F.
49, boulevard Saint-Michel
75005 Paris Tel. 43.25.83.40

Librairie de l'Université
12a, rue Nazareth
13100 Aix-en-Provence Tel. (16) 42.26.18.08

Documentation Française
165, rue Garibaldi
69003 Lyon Tel. (16) 78.63.32.23

Librairie Decitre
29, place Bellecour
69002 Lyon Tel. (16) 72.40.54.54

Librairie Sauramps
Le Triangle
34967 Montpellier Cedex 2 Tel. (16) 67.58.85.15
Tekefax: (16) 67.58.27.36

A la Sorbonne Actual
23 rue de l'Hôtel des Postes
06000 Nice Tel. (16) 93.13.77.75
Telefax: (16) 93.80.75.69

GERMANY – ALLEMAGNE
OECD Publications and Information Centre
August-Bebel-Allee 6
D-53175 Bonn Tel. (0228) 959.120
Telefax: (0228) 959.12.17

GREECE – GRÈCE
Librairie Kauffmann
Mavrokordatou 9
106 78 Athens Tel. (01) 32.55.321
Telefax: (01) 32.30.320

HONG-KONG
Swindon Book Co. Ltd.
Astoria Bldg. 3F
34 Ashley Road, Tsimshatsui
Kowloon, Hong Kong Tel. 2376.2062
Telefax: 2376.0685

HUNGARY – HONGRIE
Euro Info Service
Margitsziget, Európa Ház
1138 Budapest Tel. (1) 111.62.16
Telefax: (1) 111.60.61

ICELAND – ISLANDE
Mál Mog Menning
Laugavegi 18, Pósthólf 392
121 Reykjavik Tel. (1) 552.4240
Telefax: (1) 562.3523

INDIA – INDE
Oxford Book and Stationery Co.
Scindia House
New Delhi 110001 Tel. (11) 331.5896/5308
Telefax: (11) 332.5993

17 Park Street
Calcutta 700016 Tel. 240832

INDONESIA – INDONÉSIE
Pdii-Lipi
P.O. Box 4298
Jakarta 12042 Tel. (21) 573.34.67
Telefax: (21) 573.34.67

IRELAND – IRLANDE
Government Supplies Agency
Publications Section
4/5 Harcourt Road
Dublin 2 Tel. 661.31.11
Telefax: 475.27.60

ISRAEL
Praedicta
5 Shatner Street
P.O. Box 34030
Jerusalem 91430 Tel. (2) 52.84.90/1/2
Telefax: (2) 52.84.93

R.O.Y. International
P.O. Box 13056
Tel Aviv 61130 Tel. (3) 546 1423
Telefax: (3) 546 1442

Palestinian Authority/Middle East:
INDEX Information Services
P.O.B. 19502
Jerusalem Tel. (2) 27.12.19
Telefax: (2) 27.16.34

ITALY – ITALIE
Libreria Commissionaria Sansoni
Via Duca di Calabria 1/1
50125 Firenze Tel. (055) 64.54.15
Telefax: (055) 64.12.57

Via Bartolini 29
20155 Milano Tel. (02) 36.50.83

Editrice e Libreria Herder
Piazza Montecitorio 120
00186 Roma Tel. 679.46.28
Telefax: 678.47.51

Libreria Hoepli
Via Hoepli 5
20121 Milano Tel. (02) 86.54.46
Telefax: (02) 805.28.86

Libreria Scientifica
Dott. Lucio de Biasio 'Aeiou'
Via Coronelli, 6
20146 Milano Tel. (02) 48.95.45.52
Telefax: (02) 48.95.45.48

JAPAN – JAPON
OECD Publications and Information Centre
Landic Akasaka Building
2-3-4 Akasaka, Minato-ku
Tokyo 107 Tel. (81.3) 3586.2016
Telefax: (81.3) 3584.7929

KOREA – CORÉE
Kyobo Book Centre Co. Ltd.
P.O. Box 1658, Kwang Hwa Moon
Seoul Tel. 730.78.91
Telefax: 735.00.30

MALAYSIA – MALAISIE
University of Malaya Bookshop
University of Malaya
P.O. Box 1127, Jalan Pantai Baru
59700 Kuala Lumpur
Malaysia Tel. 756.5000/756.5425
Telefax: 756.3246

MEXICO – MEXIQUE
OECD Publications and Information Centre
Edificio INFOTEC
Av. San Fernando no. 37
Col. Toriello Guerra
Tlalpan C.P. 14050
Mexico D.F.
Tel. (525) 606 00 11 Extension 100
Fax : (525) 606 13 07

Revistas y Periodicos Internacionales S.A. de C.V.
Florencia 57 - 1004
Mexico, D.F. 06600 Tel. 207.81.00
Telefax: 208.39.79

NETHERLANDS – PAYS-BAS
SDU Uitgeverij Plantijnstraat
Externe Fondsen
Postbus 20014
2500 EA's-Gravenhage Tel. (070) 37.89.880
Voor bestellingen: Telefax: (070) 34.75.778

NEW ZEALAND NOUVELLE-ZÉLANDE
GPLegislation Services
P.O. Box 12418
Thorndon, Wellington Tel. (04) 496.5655
Telefax: (04) 496.5698

NORWAY – NORVÈGE
Narvesen Info Center – NIC
Bertrand Narvesens vei 2
P.O. Box 6125 Etterstad
0602 Oslo 6 Tel. (022) 57.33.00
Telefax: (022) 68.19.01

PAKISTAN
Mirza Book Agency
65 Shahrah Quaid-E-Azam
Lahore 54000 Tel. (42) 353.601
Telefax: (42) 231.730

PHILIPPINE – PHILIPPINES
International Booksource Center Inc.
Rm 179/920 Cityland 10 Condo Tower 2
HV dela Costa Ext cor Valero St.
Makati Metro Manila Tel. (632) 817 9676
Telefax : (632) 817 1741

POLAND – POLOGNE
Ars Polona
00-950 Warszawa
Krakowskie Przedmieácie 7 Tel. (22) 264760
Telefax : (22) 268673

PORTUGAL
Livraria Portugal
Rua do Carmo 70-74
Apart. 2681
1200 Lisboa Tel. (01) 347.49.82/5
Telefax: (01) 347.02.64

SINGAPORE – SINGAPOUR
Gower Asia Pacific Pte Ltd.
Golden Wheel Building
41, Kallang Pudding Road, No. 04-03
Singapore 1334 Tel. 741.5166
Telefax: 742.9356

SPAIN – ESPAGNE
Mundi-Prensa Libros S.A.
Castelló 37, Apartado 1223
Madrid 28001 Tel. (91) 431.33.99
Telefax: (91) 575.39.98

Mundi-Prensa Barcelona
Consell de Cent No. 391
08009 – Barcelona Tel. (93) 488.34.92
Telefax: (93) 487.76.59

Llibreria de la Generalitat
Palau Moja
Rambla dels Estudis, 118
08002 – Barcelona
(Subscripcions) Tel. (93) 318.80.12
(Publicacions) Tel. (93) 302.67.23
Telefax: (93) 412.18.54

SRI LANKA
Centre for Policy Research
c/o Colombo Agencies Ltd.
No. 300-304, Galle Road
Colombo 3 Tel. (1) 574240, 573551-2
Telefax: (1) 575394, 510711

SWEDEN – SUÈDE
CE Fritzes AB
S–106 47 Stockholm Tel. (08) 690.90.90
Telefax: (08) 20.50.21

Subscription Agency/Agence d'abonnements :
Wennergren-Williams Info AB
P.O. Box 1305
171 25 Solna Tel. (08) 705.97.50
Telefax: (08) 27.00.71

SWITZERLAND – SUISSE
Maditec S.A. (Books and Periodicals - Livres et périodiques)
Chemin des Palettes 4
Case postale 266
1020 Renens VD 1 Tel. (021) 635.08.65
Telefax: (021) 635.07.80

Librairie Payot S.A.
4, place Pépinet
CP 3212
1002 Lausanne Tel. (021) 320.25.11
Telefax: (021) 320.25.14

Librairie Unilivres
6, rue de Candolle
1205 Genève Tel. (022) 320.26.23
Telefax: (022) 329.73.18

Subscription Agency/Agence d'abonnements :
Dynapresse Marketing S.A.
38 avenue Vibert
1227 Carouge Tel. (022) 308.07.89
Telefax: (022) 308.07.99

See also – Voir aussi :
OECD Publications and Information Centre
August-Bebel-Allee 6
D-53175 Bonn (Germany) Tel. (0228) 959.120
Telefax: (0228) 959.12.17

THAILAND – THAÏLANDE
Suksit Siam Co. Ltd.
113, 115 Fuang Nakhon Rd.
Opp. Wat Rajbopith
Bangkok 10200 Tel. (662) 225.9531/2
Telefax: (662) 222.5188

TURKEY – TURQUIE
Kültür Yayinlari Is-Türk Ltd. Sti.
Atatürk Bulvari No. 191/Kat 13
Kavaklidere/Ankara Tel. 428.11.40 Ext. 2458
Dolmabahce Cad. No. 29
Besiktas/Istanbul Tel. (312) 260 7188
Telex: (312) 418 29 46

UNITED KINGDOM – ROYAUME-UNI
HMSO
Gen. enquiries Tel. (171) 873 8496
Postal orders only:
P.O. Box 276, London SW8 5DT
Personal Callers HMSO Bookshop
49 High Holborn, London WC1V 6HB
Telefax: (171) 873 8416
Branches at: Belfast, Birmingham, Bristol, Edinburgh, Manchester

UNITED STATES – ÉTATS-UNIS
OECD Publications and Information Center
2001 L Street N.W., Suite 650
Washington, D.C. 20036-4910 Tel. (202) 785.6323
Telefax: (202) 785.0350

VENEZUELA
Libreria del Este
Avda F. Miranda 52, Aptdo. 60337
Edificio Galipán
Caracas 106 Tel. 951.1705/951.2307/951.1297
Telegram: Libreste Caracas

Subscriptions to OECD periodicals may also be placed through main subscription agencies.

Les abonnements aux publications périodiques de l'OCDE peuvent être souscrits auprès des principales agences d'abonnement.

Orders and inquiries from countries where Distributors have not yet been appointed should be sent to: OECD Publications Service, 2 rue André-Pascal, 75775 Paris Cedex 16, France.

Les commandes provenant de pays où l'OCDE n'a pas encore désigné de distributeur peuvent être adressées à : OCDE, Service des Publications, 2, rue André-Pascal, 75775 Paris Cedex 16, France.

10-1995

OECD PUBLICATIONS, 2 rue André-Pascal, 75775 PARIS CEDEX 16
PRINTED IN FRANCE
(92 96 03 1) ISBN 92-64-14661-X - No. 48319 1996